The health of the oceans

The health
of the oceans

by Edward D. Goldberg

Professor of Chemistry
Scripps Institution of Oceanography,
La Jolla, California
(United States of America)

The Unesco Press   Paris 1976

Published by The Unesco Press
7 Place de Fontenoy, 75700 Paris
Printed by Imprimerie des Presses Universitaires
de France, Vendôme

ISBN 92–3–101356–4
French edition: 92–3–201356–8
Spanish edition: 92–3–301356–1

Preface

The United Nations Educational, Scientific and Cultural Organization, through its Intergovernmental Oceanographic Commission (IOC), has for several years played a leading role within the United Nations system in focusing the attention of Member States on the problems of marine pollution, especially as it affects the world ocean.

In 1973 the International Co-ordination Group for the IOC's programme, Global Investigation of Pollution in the Marine Environment, asked the secretary of the IOC to 'take the initiative in preparing a preliminary report on the health of the ocean and . . . appoint a consultant . . . to compile and analyse relevant data'. Dr Edward Goldberg was asked to undertake this task.

It is widely realized that the study of marine pollution is still in its infancy and that conflicting views are the rule rather than the exception; although Dr Goldberg has asked a number of leading scientists to review his report, the scientific opinions expressed in this work remain the author's and should not be interpreted as the views of Unesco.

A. V. Holden, an expert in pollution, both marine and terrestrial, of the Department of Agriculture and Fisheries for Scotland, kindly agreed, at the request of the secretary of the IOC, to write a foreword drawing the reader's attention to the nature of the difficulties encountered in discussing marine pollution.

The newly formed Working Committee for the Global Investigation of Pollution in the Marine Environment (which will replace the International Co-ordination Group) will be asked to keep the state of the health of the oceans under continuing review, and the leading United Nations Specialized Agencies and other relevant regional and international organizations will be asked to collaborate in this undertaking; this report is regarded as a starting point for such a review. It is hoped that it will provide those concerned with the study or control of marine pollution with a touchstone by which to judge the progress of efforts to abate or prevent it.

The author has drawn his information mainly from the international scientific literature on the subject. Consequently, the designations employed and the presentation of the material in this publication do not imply the expression of any opinion whatsoever on the part of the secretariats of Unesco and IOC concerning the legal status of any country or territory, or of its authorities, or concerning the delimitations of the frontiers of any country or territory.

Unesco gratefully acknowledges the financial support of the United Nations Environment Programme towards the publication of this report.

Foreword

Professor Goldberg was asked by the secretary of the Intergovernmental Oceanographic Commission to prepare a preliminary report on the health of the oceans. This report, which will be a starting point for a continuing review of the health of the oceans by the IOC's Working Committee for the Global Investigation of Pollution in the Marine Environment, is the result of a critical review of the literature on marine pollution. Although Professor Goldberg has an extraordinarily broad knowledge of marine pollution, he has prudently asked a number of fellow scientists to review his text, particularly where it touches on their individual specialities. Nevertheless, he was faced with a very difficult problem. In most of the ocean water, away from the places where large amounts of a pollutant may be discharged and cause locally severe pollution, the concentrations of pollutants are generally extremely low, so that measurements often have to be made near the limit of sensitivity of the analytical method used. The slightest contamination of the sample on which the measurement is made can consequently bias the results seriously. Such contamination can arise from the ship or platform from which the sample is taken, from the sampling device itself, from the person sampling, or during the subsequent analytical process, unless great care is taken.

The reader may wonder what the problem is, if the levels of certain pollutants are so very low. The main reason is that we still know very little indeed about what happens to pollutants once they have entered the ocean. They may be altered, passed into the sediments, taken up by organisms, or returned to the atmosphere, to cite a few possibilities. In other words, we know very little about the capacity of the oceans to deal with, and dispose of, man's discards. As the sea covers four-fifths of the earth's surface, and is vast in volume, it is inevitable that it will be used as a repository for waste.

It may be that, at least for certain substances, the pollutant is transferred to the sediment fairly quickly and the concentration in the water is not unduly changed. Is this a satisfactory situation, or will accumulation in the sediment lead to other problems? Can the substance be returned from the sediment to the water above? Some pollutants may increase in concentration in the water, and have little opportunity to pass to the sediments. Some chemicals may be decomposed in the ocean to harmless substances but others perhaps cannot be, or may be converted to an even more toxic form.

The rates at which these processes occur will differ widely between the pollutants, and each has to be studied separately. We should be careful about those which increase in concentration in the water, with possible serious biological effects in the long term. Professor Goldberg is extremely concerned about our reaction times, not only to well-defined pollution dangers, such as high-level localized discharges, but also to warning signs provided by small, slow but continuing increases in concentrations. Thus, the doubling of the concentration of, say, lead in open-ocean waters may be a significant and serious development, even though the concentration is very low. Because fish can accumulate some substances to concentrations many times higher than in the surrounding water, low concentrations cannot be ignored.

Pollutants may enter the oceans by dumping, spillage, discharge from factories or cities either directly or via rivers, or perhaps be transported through the atmosphere and washed in by rain. Many substances enter the oceans from natural sources as well as by man's activities, and it is important to know whether we are making a significant contribution to the natural input. The necessary calculations are difficult, but Professor Goldberg has made an attempt to assess man's influence on the rates at which several naturally occurring substances enter the sea.

Many estimates of the input of pollutants to, or their presence in, the oceans depend on the measurement of very small concentrations. Some of the analytical methods used to measure them are near the limit of sensitivity. There may be no agreed international reference standard for the substance concerned, and no intercalibration of methods between analysts. This leads us to one of the greatest difficulties in discussing marine pollution: an uncertainty about the accuracy and precision of many of the measurements of marine pollution concentrations. As an example, following the discovery in 1966 that polychlorinated biphenyls were being detected during the gas chromatographic analysis of DDT and its immediate derivatives (DDE, DDD), and were interfering with the interpretation of the DDT data, it was realized that many measurements of DDT prior to 1966 were suspect and therefore unreliable. At the low concentrations believed to be present in sea water, other substances may also interfere in the analysis. Perhaps the most accurate estimates of pollutants are those of the radio-active isotopes, for which highly sophisticated equipment has been developed over the last thirty years.

No wonder then that the subject of marine pollution is polemical, and we must congratulate Professor Goldberg for having the courage to attempt this assessment on the basis of the best available information, mixing as hard fact as possible with careful speculation to see in which direction we appear to be going, and sometimes coming up with comfortably similar answers from differing premises.

Generally speaking, Professor Goldberg's reviewers have welcomed his ap-proach even while expressing some misgivings about some of the assumptions he has inevitably had to make, or questioning some of the values he has drawn from the literature for his calculations. This is a natural consequence of the still comparatively rudimentary state of the art of marine-pollution measurement and assessment. While improvements will surely come in the accuracy of the necessary measurements, the deductions made regarding the likely biological effects of the pollutant levels found are likely to be the basis of controversy for a long time. Yet we must not limit our considerations to the present, but have regard to the health of the oceans for the benefit of future generations. The reader should therefore keep this in mind when reading this book; he should, as it were, see the forest rather than look at the trees.

A. V. HOLDEN

Contents

Introduction

The question as to why another book on marine pollution has confounded the author as it most probably will the reader. The most appealing rationale to the author is that many of the existing volumes are dedicated to describing the present situation rather than to attacking the more formidable problem of asking what we should know, in relation to what we do know, about the alteration of the ocean's chemistry by man, in order to regulate the releases of materials to the ocean. It is in this latter mood that the publication has been prepared.

A rather unusual situation exists in marine-pollution problems. Most of the already identified pollutants that may jeopardize marine resources are in extremely low concentrations, both in sea water and in marine organisms, levels difficult to assay by competent chemists. As a consequence, there are very few analysts in the world capable of carrying out reliable analyses.

Thus, the small amounts of data produced by the few chemists come under intense assessment. There are less than a dozen laboratories in the world capable of measuring DDT or petroleum components in sea water. There are perhaps a similar number of laboratories that are engaged in analysis of transuranic element concentrations in the ocean system. On the other hand, the more readily measurable heavy metals, some of which are classified as pollutants, enjoy an enormous literature. Thus, in one way the task of producing this book was simple, especially when focusing upon the more difficult-to-measure substances. Problems arose when trying to filter the substantial and significant information on heavy metal pollutants from the many numbers about their levels in marine samples, a large proportion of which appear to be random.

Many of the author's colleagues have reviewed the individual chapters and to them he is most grateful. The final product represents the author's views and not necessarily those of the reviewers, who included Daniel P. Serwer of the United Nations Environmental Programme (UNEP), Geneva; Thomas W. Duke, of the United States EPA Gulf Breeze Environment Research Laboratory; V. Zitko of the Environment Canada Biological Station at St Andrews, New Brunswick; A. V. Holden of the Department of Agriculture and Fisheries for Scotland, Freshwater Fisheries Laboratory, at Pitlochry; C. S. Giam of the Texas A & M University College of Chemistry; M. Ehrhardt of the Institut für Meereskunde an der Universität Kiel; John W. Farrington of the Woods Hole Oceanographic Institution; Jun Ui of the University of Tokyo and G. G. Polikarpov of the Academy of Sciences, Ukraine S.S.R. Institute of Biology of the Southern Seas.

1.   Oceanic and societal time scales

Many thousands of substances enter the oceans as a consequence of material usage and energy production by human society. Some, such as the pesticides DDT and dieldrin and the artificial radio-active materials produced in nuclear reactors, are alien to the marine system. Others already exist in natural waters but their concentrations are altered by man's activities. The concentration of lead in coastal waters appears to have been increased by the entry of the anti-knock agents lead tetra-ethyl and lead tetra-methyl and the products of their combustion in automobiles.

Only a few of these many compounds are likely to produce unwanted consequences such as the tainting of seafoods, changing the structure of communities of marine organisms or the loss or restricted use of non-living resources such as recreational areas. What tactics could be employed to counteract destructive impacts? Is there time to

use them effectively once formulated? Must we await a tragedy to give us an awareness of the effects of toxic substances promiscuously introduced to the oceans? Such was the case with the entry of mercury wastes to the Minamata Bay in Japan and the consequent neurological disease among the people from the consumption of fish and shellfish contaminated by methyl mercury. Or can we effectively limit the discharges of toxic materials to the oceans, as we have done with the radio-active wastes produced in nuclear-power reactors, before a tragedy occurs? The discharge of these radio-active substances to the seas has been regulated upon the basis of the best available scientific information provided to those responsible for the management of marine areas. As a consequence, their oceanic concentrations have been maintained at levels that provide no evident hazard to living beings.

One factor that threads its way through

these questions is time. How do the time scales of response and action by governmental groups compare with those of natural processes? What periods of time are involved in the development of a catastrophe in coastal zones resulting from the discharge of harmful substances? How long does it take for scientists to recognize and define a real or potential threat to the marine environment after the uncontrolled release of a pollutant material?

I shall attempt to answer these questions using recent case histories as my guide. Initially I shall consider the times materials spend in the sea-water system between their introduction and their assimilation or their chemical breakdown. Then, I shall examine some recent examples of governmental responses to scientific advice and to catastrophic events and the reaction times involved. I shall look at the economic impediments to effective marine resource management. Finally, I shall pose some potential future problems and some possible tactics for their solution.

Oceanic time scales

The times that chemicals spend in sea water or other geological domains may be estimated by means of simple mathematical models in which the oceans are depicted as a vast reservoir for continental materials mobilized by winds, rivers and glaciers. Substances taking part in these natural weathering processes are usually considered to have time invariant concentrations in the ocean waters as well as in such other reservoirs as the atmosphere and biosphere.[1] On this basis the amount of a material introduced to the oceans in any given time period is compensated by the loss of like amounts precipitating to the sediments or undergoing destruction through such processes as radio-active decay or microbial activity. Hence, the time a chemical spends in the

ocean water (or in some other reservoir such as the atmosphere or biosphere) is given by the total amount in the oceans (or in the atmosphere or biosphere) divided by the total flux into or out of the ocean water (or atmosphere or biosphere). This period is known as the 'residence time'. The simplest schematization is given in Figure 1 where the world ocean is considered as a single reservoir for continentally derived materials. More complicated models have been devised (see Chapter 2). Some divide the world into parts based upon the major ocean basins—the north Pacific, the south Atlantic, Indian, etc. Further, each basin may be subdivided into two components, a mixed and a deep layer. The mixed layer corresponds approximately to the zone in which the primary production of organic matter takes place through photosynthesis. In all of these models the assumption is made that there is a complete mixing of the substance within a reservoir in a relatively short time compared to its residence time.

Several characteristics of these models are important in pollution studies. First of all, the time required for a substance to achieve a steady-state (i.e. time-invariant) concentration in a reservoir is given by a period of four residence times. Conversely, for a pollutant whose supply to a reservoir is cut off after a steady-state condition is reached, it will take a period of four residence times for its concentration to fall off to less than 1 per cent of its steady-state value.

Secondly, the residence time of a chemical is inversely related to its reactivity in the reservoir, so that persistent (low reactivity) chemicals will have longer residence times, whereas the more reactive chemicals will be removed more quickly

1. The term 'biosphere' as used in this publication is defined as the part of the terrestrial environment actually occupied by living or recently dead organisms.

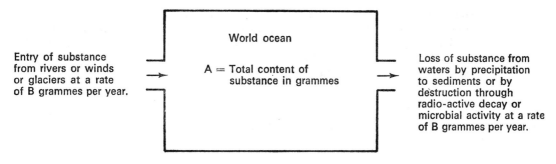

Residence time = A/B

Fig. 1. Simple schematization of a steady-state world ocean.

either by breakdown or by precipitation to the sediments.

A coupling of these two concepts may allow the prediction of pollutant levels in the ocean waters where the residence time of a naturally occurring substance or of a previously studied pollutant with similar chemical properties is known.

The coastal ocean

The world ocean can be conveniently divided into two zones for the purposes of both scientific and societal inquiries: the coastal ocean and the open ocean. The coastal zone constitutes about 10 per cent of the total oceanic area and includes estuaries, lagoons, inshore waters and many marginal seas and waters over the continental shelves and slopes. The North Sea, Chesapeake Bay, the Persian Gulf and the Sea of Japan are examples of coastal waters. Their properties are strongly influenced by boundaries with the continents and with the sea floor. The coastal ocean receives direct injections of continental materials via rivers, direct terrestrial runoff and drainage, and the atmosphere, and through such mobilizing agents of man as domestic and industrial outfalls and ships.

These coastal waters are the sites of high biological activity. The marine primary production of organic material, the photosynthetic formation of plants which forms the base of the food chain that ends in fish, birds and marine mammals, takes place predominantly in these waters. The open-ocean areas, with a few exceptions such as some productive equatorial waters, are the marine deserts. Within the coastal zone there are some especially productive waters—upwelling areas—where a coupling of strong offshore winds with prevailing boundary currents brings nutrient-rich deep waters to the surface. Here there are even higher levels of primary productivity with the consequent production of large stocks of fish. The most important upwelling areas are off Peru, California and the western coasts of Africa and are estimated to constitute only 0.1 per cent of oceanic areas; yet they are responsible for about 50 per cent of fish production.

The life in the coastal sea strongly influences the fates of materials introduced from the continents. Some organisms have a remarkable ability to accumulate substances from sea water, even where the materials have extremely low concentrations, say in the parts per billion[1] or parts per trillion range. For example, vanadium existing in sea water at levels of about one part per billion is enriched in the blood of tunicates to a level of parts per thousand. DDT and its residues are found in surface sea waters in concentrations of parts per trillion, yet in the fish levels of parts or tens of parts per million are not uncommon (see Chapter 3). As a consequence of such abilities, the organisms of the sea may act as conveyors of man's wastes.

If a substance becomes incorporated in skeletal or other persistent parts of an organism, it is likely to be carried to the sea floor upon death of the organism. If it becomes incorporated into soft tissues, it may be recirculated upon death. Substances accumulated in living organisms may return to man in the form of food. What periods of times are involved in such processes?

The persistence of chemicals in coastal waters, before removal by sedimentation, by decomposition, by mixing with the open ocean, or by the harvesting of living organisms, spans months to years. Estuaries exchange their waters with the open ocean in such periods. Recent work at the Scripps Institution of Oceanography showed that the residence times of radium and lead isotopes in the highly productive Gulf of California waters were a few months and about a month, respectively. Calculations of this type assume a steady-state situation in which the concentration of the substance in the water is invariant with time—that is, the amounts that enter the system are compensated by like amounts that are removed by one process or another. For most pollution problems, however, there is a gradual build-up of the substances of concern.

The time from initial entry of a pollutant to an awareness of a problem may span decades. The Minamata Bay incident started in the late 1930s when the Chisso Corporation, one of the leading chemical industries in Japan, began its production of vinyl chlorides and formaldehyde at its factory on the shores of Minamata Bay. Spent catalysts containing mercury were discharged into the bay. The fish and shellfish accumulated the mercury in the form of methyl mercury chloride. Their consumption by the inhabitants of the area, primarily the fishermen and their families, resulted in an epidemic of neurological poisoning. The first occurrence of the disease was recorded in 1956, over 15 years after the first entries of wastes to the bay. It was not until 1959 that mercury was associated with these afflictions. During the preceding ten years the production of acetaldehyde, one of the prime users of mercury, increased by over four times. The time between the initiation and the detection of the problem, in this case, was somewhat over a decade.

A slightly longer period is associated with the detection of the reproductive failures in the brown pelican colony on Anacapa Island off the California coast from 1969 to 1972, allegedly resulting from extensive DDT pollution. The DDT and its degradation products were accumulated in the marine organisms that eventually became food for the birds. The result was the production of thin-shelled eggs which broke easily. The Montrose Company had twenty-five years previously started the manufacture of DDT with the concurrent discharge of wastes into a sewer system and subsequently into the sea. The DDT first appeared in the sediments of the southern California area in 1952 with progressively increasing concentrations occurring up to 1969.

---

1. In this book, the term billion follows American usage and is equivalent to 1,000 million. Therefore, the term trillion is equivalent to 1,000 billion.

The open ocean

The open ocean differs significantly from its coastal counterpart not only in its time scales but also in its relationships with the continents and sediments. The time spans describing natural processes extend from hundreds to hundreds of millions of years as contrasted to months to decades in the coastal ocean. For example, aluminium species, entering the oceans as a result of weathering processes on the continents, spend about a century in solution before precipitating to the sediments. In comparison, sodium, one of the principal elements in sea water, resides for perhaps 100 million years in the open-ocean water before accommodation in the sediments.

The open ocean constitutes 90 per cent of the area of the total world ocean. Its deeper waters, those generally considered to be 100 metres or more below the sea surface, are out of contact with coastal waters or surface waters for periods averaging between a few hundred and a thousand years, depending upon the specific ocean basin. Herein lies not only the value of these waters as acceptors of some of man's wastes today but also the perils for the future that can arise from small, but continuous, introductions of highly toxic substances.

For example, open-ocean waters contain around 100 million tons of mercury, an amount vastly greater than the 10,000 tons mined annually by man.[1] The deep open ocean can accept low-level mercury wastes, especially in the inorganic forms (see Chapter 5). Similar arguments can be made for the disposal of such toxic metals as arsenic and antimony.

On the other hand, the long residence times of chemicals in the open ocean may lead to the formation of a toxic broth through the slow accumulation of man's wastes. There is taking place today in the deep waters of the open ocean a gradual, but continuous, build-up of synthetic organic chemicals containing chlorine, and sometimes fluorine, atoms, the so-called halocarbons. Initially entering the coastal ocean through sewer outfalls and the surface waters of the open ocean via the atmosphere as gases, they are transferred by physical and biological processes to the deep ocean in less than a decade. Some of these chemicals are known to interfere with metabolic processes of living organisms (see Chapter 3). The heavier compounds like DDT and its degradation products or the polychlorinated biphenyls (the PCBs) affect the calcium metabolism of marine birds with the result that their eggs have thin shells. The lower-molecular weight species, such as the chlorofluorocarbons, which are used as aerosol propellants, may interfere with the fermentation activities of microorganisms. As a possible consequence, the degradation of organic matter in the normal biochemical cycles would be inhibited. In addition man's use of fermentation, such as in sewage-digestion systems and in the production of wines and beers, could be affected adversely.

Our concern is the haunting possibility that levels of a toxic material can reach such values that exposures of organisms to such materials in the open ocean, as well as in the coastal ocean, result in widespread mortalities or morbidities. For at such a time there is no turning back. The great volume of the open ocean makes removal of a toxic substance, identified by a catastrophic event, an endeavour beyond mankind's capabilities with the technologies of today or of the foreseeable future.

The sediments of the open ocean accumulate natural debris at a rate a thousand or so times slower than coastal deposits. The fluxes of solids, mobilized during weathering processes and carried to

1. The term 'ton' as used in this book means metric ton.

the open ocean primarily through the atmosphere, are much smaller than those entering coastal waters from rivers and from direct terrestrial runoff. Man's wastes will also be accumulated much more slowly in the open ocean. Sewer outfalls and ships will enhance the contributions going to coastal waters.

The characteristics of the open ocean determine in part the ability of its sediments to accommodate the wastes of man. Where the vast areas are attractive for such a purpose, the slow accumulation rates allow prolonged contact of the sediment components with overlying waters where dissolution processes can take place. Still, any toxic substances going back into solution are removed from surface and coastal waters for centuries.

Since biological activity is less intense in open-ocean surface waters, there is a smaller potential for the downward transport of pollutants through biological activities. Most of the materials dispersed to the open-ocean environment through man's activities are still in the water column. Only a very few have been taken up by the sediments.

The open-ocean water column carries in it today many signatures of our technological society. Radio-active isotopes of strontium and cesium, produced primarily in nuclear-bomb detonations, have been found to depths about one-fourth of the way down to the sea floor (to depths of 1,000 metres or so). Other radio-active isotopes, such as those of cerium and promethium, are chemically more reactive and have been found at even greater depths, although in smaller concentrations than those in surface waters. The pesticide DDT and its degradation products have been found in all open-ocean organisms analysed over the past five years.

The surface of the open ocean is soiled with petroleum products in the forms of tar balls or of coatings (oil slicks) whose thicknesses have molecular dimensions. In addition, ocean currents carry litter, plastics, glass, wood products and metals, many of which are used to contain products of commerce. These visible alterations of the surface signal the need for continual measurement and assessment of the invisible pollutants lying below them.

Societal time scales

Governing bodies are alerted to regulating the environmental releases of substances that can jeopardize marine resources by the predictions and advice of the scientific community or by a catastrophic event. What have been the recent experiences with response times of society to either of these two triggering actions?

The possibility that the discharge of high levels of radio-activity from nuclear-power reactors or from nuclear detonations could bring about undesirable consequences to public health and to the vitality of marine organisms prompted scientists to action in the years following the Second World War. The primary concern was the return of ionizing radiation to man either in the form of food, through direct exposure by swimming or through recreational activities on beaches (see Chapter 4). There was the sense that it would be convenient, and perhaps necessary, to dispose of some amounts of radio-active nuclides deliberately to the seas. Did we have an adequate scientific basis for ascertaining what amounts the oceans could accommodate safely? The complexities of oceanic chemical processes did not permit the precise prediction of the fate of a specific chemical introduced to a particular place. There were great areas of ignorance in our understanding of oceanic phenomena. Hence, the initial recommendations for sea disposal of radio-active wastes contained an explicit sense of caution, of control and of limitation. They were intended to be experimental. During the following years, the amounts of radio-activity introduced to the oceans have been strictly limited under national regulations which have been based upon internationally promulgated standards (primarily by the International Atomic Energy Agency (IAEA) and the International Commission on Radiological Protection (ICRP)).

Guidelines for radio-active waste disposal today are based upon the concept that man is one of the most sensitive organisms to ionizing radiation. As a consequence of cell destruction or alteration, two types of effects are possible: somatic and genetic. The former encompasses the mortalities and morbidities associated with either acute or chronic exposures. Such diseases as cancer or leukemia can result, especially where heavy dosages have been received. Our knowledge of illnesses due to long-term exposures to low levels of radiation is much less fully developed than our knowledge of the effects of high-level exposure. A similar situation exists for essentially all other pollutants. Genetic effects are mainly due to chromosomal damage; defects may be observed in individuals many generations removed from the victim of exposure.

The policies of the United Kingdom constitute an example for the national management of radio-active materials. Minimum risk to its citizenry is sought on the grounds of both somatic and genetic effects. Of primary importance is the identification of potentially significant pathways back to man for radio-active substances released to the environment. Usually these pathways involve the ingestion of foods or the inhalation of atmospheric constituents. Also, for any specific isotope they are few in number, and in many cases only one has been found to exist. Once identified, they provide the basis for the 'critical pathways approach', a management strategy which provides an adequate and economic surveillance of the return routes (see Chapter 9).

Given the need to deal with the potential releases of highly toxic substances with inadequate information to provide precise answers to important questions, the conservative approaches described above, suggested by scientists and adopted by autonomous nations, have proven successful. Legislation on a national basis has developed over periods of years. For example, in the United Kingdom, the Ministry of Agriculture, Fisheries and Food is responsible for maintaining the integrity of coastal waters from undesirable impacts of radioactivity released in the coolant waters of nuclear-power reactors. Permissible levels for radionuclides in sea foods, for example, cesium isotopes in fish and ruthenium isotopes in algae, have been adopted. On these bases and on the results of a continual monitoring of levels in the organisms and in the reactor effluents, the return of radiation to man has been reduced to a level considered as an acceptable risk to his health.

How quickly can governments respond to an unexpected catastrophe resulting from the unregulated discharge of an unknown toxic substance into the marine environment? The Minamata Bay incident in Japan and subsequent events provide a time scale that may better reflect a unique socio-economic framework than a general set of reaction times which could be expected in any similar happenings elsewhere in the future. Still, it is worthy of review for it may provide us with a guide as to how more rapid responses could be achieved.

Following the observation of the disease in 1956, primarily among fish-eating fishermen and their families, it took about three years to ascertain that mercury was in high concentration in the fish of the area and in the dead patients. About eighty cases of neurological disorders, later recognized as Minamata Bay Disease, were diagnosed up to 1959. Articles in environmental and medical journals throughout the world began to appear in 1960 describing both the disease and the circumstances under which it developed. At this time the prevailing mood was that a still unidentified organic compound of mercury was the culprit. The Chisso Corporation in these early periods was making every effort to avoid an association with, or a responsibility for, the disease. The company had gathered together a group of scientists who refuted the evidence of the workers at the Kumamoto Prefecture University. These latter investigators had established the relationship between mercury and the disease. The company then found new discharge sites for its wastes in a northern area and this was followed soon after by the discovery of several new cases of the disease there. At this time, the citizens of Minamata Bay had become aware that the local sea food was the source of the epidemic. There developed strong animosities between the fishermen and the Chisso Corporation, as the demand for fish by the local inhabitants fell to zero. The fishermen became violent and stormed and destroyed the offices of the company. Through such actions the Minamata Bay Disease became known throughout Japan. Very small compensations were paid by the company to the fishermen for the loss of their livelihood and to the victims of the disease. For a few years the problem was forgotten. However, in 1965 a second outbreak took place in a new location along the Agano river in Niigata. Here the source of the pollution was attributed to an acetaldehyde factory of the Show Denko Company, discharging its spent mercury catalysts into the river. The patients of the second disease initiated a civil action in the court in June 1967; this is presumed to be the first large civil suit brought against a polluter in the history of modern Japan.

A judgement against the Show Denko's plant for their promiscuous release of mercury wastes to the Agano river was handed down by the Niigata District Court in September 1971. Compensation of around $800,000 was allotted to the seventy-seven victims or their families.

In March 1973, the Kumamoto District Court decided that the Chisso Corporation was at fault in the discharge of mercury wastes to Minamata Bay. Payment of about $3.8 million to forty-five victims or their families was ordered by the court.

In 1963, seven years after the diagnosis of the first cases, the active agent causing the disease was identified as methyl mercury chloride. This introduced a new dimension into the marine chemistry of mercury. Before this discovery it was thought that the only forms of mercury involved in natural processes were inorganic. But further surprises were in the offing. In Sweden, as a consequence of some disastrous impacts of mercurial pesticides upon wildlife, there was an active group of scientists working on the environmental chemistry of mercury. Their analyses of mercury in uncontaminated fish indicated that nearly all of it was in the form of an organic compound, methyl mercury. The toxic form causing the Minamata Bay Disease was similar to the naturally occurring form. These results were eventually confirmed by Japanese scientists.

The first group to systematically evaluate the risks in the consumption of fish containing mercury, at either naturally occurring levels or at levels enhanced by man's activities, was appointed in 1968 by the Swedish National Institute of Public Health in conjunction with the Swedish National Board of Health and the Swedish National Veterinary Board. The group assessed the toxicological evidence from the Japanese epidemic and the fish-eating habits of both the Japanese and Scandinavian populations. It was found that in Sweden there was a small number of persons without symptoms of Minamata Disease who had consumed fish to such an extent that mercury concentrations in their hair and blood were the same as those of Japanese who had shown neurological symptoms of poisoning. These results emphasized the varying sensitivity of people to methyl mercury poisoning and the varying eating patterns of countries.

Although the group was not charged with the formulation of allowable methyl mercury levels in fish, they did consider the problem in their report published in 1971. The Swedish population consumes on an average 56 grammes of fish per day per person. With a level of 0.5 p.p.m. of mercury in the fish, 10 per cent of the people might carry the maximum tolerable level of mercury in their bodies, i.e. an amount which as yet has not produced symptoms of Minamata Bay Disease. The high consumers of fish might build up in their bodies amounts that could lead to the disease. With a level of 0.2 p.p.m. in the fish, all exposures would lead to acceptable body levels.

Subsequently the 0.5 p.p.m. limit was adopted by Sweden, followed by several other countries including the United States where the average fish consumption is 17 grammes per day, less than a third of that of Sweden. These limits pertain not only to coastal fish, which may have had their mercury levels elevated by man's discards of mercury to marine areas, but also to deep-sea fish such as tuna whose mercury levels appear to have remained unchanged over the past century.

In retrospect, perhaps the rationale for the 0.5 p.p.m. level needs re-evaluation by the scientific community. Mercury, at these concentrations, has existed in deep-sea fish for many centuries, well before extensive mobilizations by man (see Chapter 5). The swordfish and the tuna, both deep-sea fish, simply have not added mercury to their bodies as a consequence of man's activities. If these fish do provide a health hazard to those who eat them frequently, their consumption should be regulated on the basis of a natural toxin, the methyl mercury. But

more extensive toxicological studies are needed to reach such a decision.

It took over two decades for the Japanese Government to halt the discharge of mercury into the coastal marine zone and for other countries to define what are acceptable mercury levels in sea foods. Recently, mercury has become one of the blacklisted substances in the ocean-dumping convention.

The Minamata episode has shown that scientists can reach an understanding of a critical pollution problem in the coastal zone and can propose effective courses of remedial actions in times of decades, the periods over which the problem developed. Similarly, the increased levels in fish from the Baltic Sea adjacent to Stockholm were reduced when the Swedish Government, acting upon advice of their scientists, halted the discharge of mercury wastes from chemical plants. Marked reductions were evident in periods of under a decade, following constraints upon the promiscuous release of mercury to the environment. However, such short times of action/reaction are not applicable to problems that might develop in the open ocean, where time periods of reducing oceanic pollutant concentrations will be hundreds to thousands of times longer than those in the coastal zone.

As a final case study, it is desirable to consider the numerous attempts to control the releases of petroleum to the oceans. Such actions are generally stimulated neither by mass mortalities nor by scientific concerns but by the aesthetic insults that have resulted from these releases. One of the early activities involved the United States Government which invited thirteen countries to Washington, D.C., in 1926 to prepare a convention on the prevention of oil pollution. Since that time a large number of international groups have attempted to minimize the oil pollution of the seas by means of conventions.

The management of oil has been clearly less successful than the management of either radio-activity or of mercury. I suspect that this primarily results from the threats to human health of the latter substances with little evidence that there is a need for such a concern with petroleum. There was an uproar over the soiling of beaches and the killing of birds following the break-up of the tanker *Torrey Canyon*, and with the blow-out of the oil well at Santa Barbara. There would probably have been a much greater agitation on the part of the public and action on the part of politicians had there been evidence that humans were killed or maimed by inadvertent contact with petroleum. Public health has greater power than dead birds in instigating legislation or conventions.

Societal restraints

There are several impediments, each with an economic root, to the development of a scientific basis for the description and for the forecasting of ocean-pollution problems. The first involves the difficulties in acquiring a knowledge of past, present and predicted production, use and disposal data for chemicals that may insult the marine system. The second, and perhaps more worrisome, is associated with the staggering development of the multinational corporations (MNCs) whose manufacturing activities compete with those of great nations in the utilization of materials and energy. Effective controls upon their actions have not as yet been formulated by international organizations, although debate on the subject has now started in the United Nations. Yet their concern for the environment in the prosecution of their activities appears to be minimal.

A continual assessment by scientists of the production and uses of chemicals and energy can provide a means to predict

which substances, if released to the environment, might jeopardize the continued uses of marine resources. Such evaluations usually involve estimates of their toxicity and the construction of mass-balance models—schematizations to understand the flow patterns of materials from the site of release to their environmental reservoirs. However, there are obstacles to such undertakings, for the production, use and disposal data are often maintained as privileged information by manufacturers or by sovereign nations. For example, in the United States, when a chemical is produced by no more than two companies, the production and use figures are proprietary information. The rationale that prohibits the release of such data is economic. Governments have an obligation to protect their manufacturers from being placed at an economic disadvantage through the publication of production and use data. Yet it is in the interest of their citizenry to maintain a continuing assessment of the state of their marine resources, for which such data are essential.

One of the first recognitions of this dilemma took place in 1970 at a National Academy of Sciences Workshop on halogenated hydrocarbons in the marine environment at which the impacts of the polychlorinated biphenyls were assessed. The sole United States manufacturer had refused to release its production figures, although requested to do so by many scientists and government officials. For another year, the scientists were frustrated in their quests for this data. However, pressures by concerned scientists and laymen eventually caused their release. The economic consequences of this action have not been disclosed by the Monsanto Company, the involved United States producer.

The need for such data on both national and international levels has constituted some of the background noise at meetings of environmental scientists. New concerns are evolving for which production and use figures are urgently needed. For example, there is a geographical shift in the applications of DDT in agricultural and public-health usages (see Chapter 3). Initially, most of the DDT was employed at mid-latitudes of the northern hemisphere. It now appears that the centre of usage is shifting southward as bans upon DDT are imposed in industrial countries. However, the exact details of this change are not available to the scientists to assess the possible impact upon marine life. There are other examples of frustration of scientists along this line. Details of the discharges of trans-uranics from nuclear reactors and reprocessing plants are wanted by a concerned group of marine scientists. For military and economic reasons, much of the data is classified.

As a consequence, the Intergovernmental Oceanographic Commission early in 1974 formed an *ad hoc* group of experts (POOL—Pollution of the Ocean Originating on Land) who have been directed 'to suggest practicable and effective means for obtaining information on the quantities of important pollutants, present and potential, introduced into the ocean from land-based sources by whatever route'. The charge is well defined—the task appears formidable.

Of greater concern than either the manufacturer of a chemical or the sovereign nation are the multinational corporations (MNCs) which have emerged as dominant global institutions. Their potential influences can be seen in comparison of their annual products with those of the leading industrial countries. For example, General Motors, with an annual product of $35.8 billion in 1973, exceeded the gross national product of Switzerland ($20 billion) and Denmark ($15.5 billion). Ford Motors ($23 billion) and Exxon ($25.7 billion) were slightly ahead of Austria ($14 billion). Eleven MNCs in 1973 had annual gross

products in excess of $10 billion. But what is more foreboding is the observation that their annual rate of growth exceeds that of national States. Most probably there will be more transnational mergers of firms and a consequential increase in the number of MNCs.

The MNCs seek lower internal and external costs of operation. As a consequence there is a general drift in their activities from the mid-latitudes of the northern hemisphere (the developed nations) southward. In the developing nations the labour costs are usually lower and often the concern for the environment is overridden by their need for economic development. Over-all, the MNCs do tend to improve the living standards of the world in general and the emerging nations in particular, but often with a blindness toward their impact upon surroundings.

There are many instances of this industrial drift to the south, mediated by the MNCs. For example, the dramatic development on the coastal areas of Portugal of petrochemical industries, which will supply northern Europe, is sponsored by MNCs. An indication of the importance of this operation may be seen in the construction of the world's only drydock capable of handling million-ton tankers in Lisbon. Chemical industries have been transferred from the United States to Puerto Rico. It is important to know whether the internal costs of labour or the lesser concerns about the environment dominated such developments. Clearly, one of the problems facing the international community of nations is to devise strategies of dealing with the MNCs especially as they impinge upon common environmental resources. The weak link in plans to protect the world ocean may be the country willing to negotiate the quality of its coastal environment for short-term economic gains.

Some problems of tomorrow

The time it takes a pollutant to reach an unacceptable level in sea waters is generally equivalent to the time it will take to fall back to an acceptable level if the supply is cut off. A pollutant will reach steady-state levels in time periods equal to about four times its residence time and will return to original conditions, upon cessation of release, in the same period. In the case of lead introduced to biologically productive coastal waters, the residence time of a month means that in four months maximum (steady rate) concentrations will be reached assuming continuous and uniform injections to the environment. If the pollutant lead supply were cut off, the waters would probably approach the pre-historic lead levels in about four months. For DDT and some radio-active species entering coastal waters the times involved may be years or decades. Responses by nation-States to limit releases, following catastrophes or assessments by scientists, have usually taken place over a period of a decade or less.

Another set of problems exists, on a longer time span, whose resolution may require the concerted actions of groups of countries. Such problems involve the very small leakages of highly toxic substances to the oceans from many locations in many countries. Build-ups to unacceptable levels may take centuries. Examples include the transuranics and some halogenated hydrocarbons. Let us consider the former in some detail. The transuranic elements include those metals heavier than uranium in the periodic table (see Chapter 4). They are produced in nuclear-power plants and in atomic weapons, either on standing or during utilization. The nuclear fuels in both the weapons and the reactors must be reprocessed periodically to remove the transuranics that have built up over time as a result of nuclear reactions. The interval between successive reprocessings appears to be from one to five years. It has been estimated that in 1980 there will be about 100 tons of plutonium (one of the transuranics) in United States reactors and weapons. Since reprocessing sites are located away from use sites, a movement about the United States of between 20 and 100 tons of plutonium per year is envisaged.

The inhalation of microgramme (10^{-6} g) quantities of plutonium may lead to the development of lung carcinomas in humans. This is the basis of the plutonium or transuranic problem. Can we contain the plutonium (and other transuranics) during production, transport and reprocessing such that there is a minimal possibility of the inhalation of microgramme amounts of these substances where there may be 10 billion times this amount in circulation!

Upon dissemination to the environment, much of the plutonium (and other transuranics) may enter coastal marine waters where it is known to be concentrated by algae. There are any number of orchestrations of events that could lead to a return to man through inhalation (see Chapter 8). To minimize the risk to public health during the oncoming nuclear age, we must be able to predict the possibility of such a happening. For such purposes we need production, use and disposal data from all countries involved in the production of transuranics, and the locations of use, disposal, production and reprocessing. In addition, we need effective surveillance systems.

A similar concern involves the halogenated hydrocarbons and halocarbons—the pesticides like DDT and dieldrin, the industrial chemicals like the polychlorinated biphenyls, the dry-cleaning solutions like perchloroethylene, the solvents like dichloroethylene and the aerosol propellants, including the chlorofluorocarbons. These materials are entering the environment in quantities up to a million tons annually (see

Chapter 3). They are widely disposed since they are atmospherically transported. They are detectable in surface ocean waters. Some interfere with life processes such as fermentation when their concentrations reach critical levels. Who is responsible for the global book-keeping on these materials such that we can turn off the release valves if danger is imminent?

The strategies for the management of the transuranics and for the halogenated hydrocarbons must be formulated by the prime producers today. One cannot expect a large collective of nations, such as those of the United Nations and its family, to address themselves to these problems which plague the technologically advanced when there are basic survival problems for the populations of developing member countries. I suggest that bilateral and multilateral recognition and agreements to attack these problems will yield productive actions. The recent agreements between the United States and Canada to protect the Great Lakes, among the North Sea countries through the International Council for the Exploration of the Sea (ICES) to understand their pollution problems, and between the United States and the Soviet Union to exchange scientists and to publish jointly a periodical dedicated to marine-quality problems are examples of necessary steps to define potentially dangerous leakages to the environment and to propose remedial actions. The efforts to solve the first-order problems of marine pollution by large international organizations have in general been both noble and futile. (The necessary information for the statement and strategies for the solution of problems has usually been unavailable.) Economic and social concerns have overridden environmental ones. The starting point is acknowledgement of a problem by the political leaders and by the scientists in concert.

2.

Marine pollution dynamics

An understanding of the present distribution and abundance of pollutants in the ocean and the prediction of future distribution and abundance demands a knowledge of the routes, reservoirs and reactions of the substance under consideration. For each pollutant there are many possible behaviours and interactions with the living and non-living components of the marine environment. Methods of identifying the significant factors that govern environmental concentrations of pollutants will be considered in this chapter.

First, we shall consider the major transport paths of materials from the continents to the oceans. There are three important natural mobilizing agencies—rivers, winds and glaciers. To these, man has added domestic and industrial outfalls and ships.

Secondly, we shall compare the fluxes of natural and pollutant materials along these routes. It will be seen that the flux of materials into the environment by the activities of man is approximately 10 per cent of the natural material flux.

Thirdly, we shall examine the steps to be taken in the formulation of mass-balance models, which attempt to describe present-day distributions and flows of pollutants through the environment. The importance of production and use data, the recognition of the chemical and physical forms of the pollutant, the identification of reservoirs in the system under study, and the methods of flux calculations will be considered. Three examples of models will be presented.

Finally, we shall consider techniques to reconstruct the history of the polluting activities of man in order to relate present and past oceanic concentrations to present and past fluxes.

Transport paths

RIVERS

The prime mover of materials from the continents to the oceans is the system of rivers (Table 1). Although estimates vary as to the total flow and to its average composition, fluvial transportation of natural materials appears to be about ten times greater than that of glaciers and about a hundred times greater than that of the winds. The particulate load of rivers is estimated to be about four to five times the dissolved load. Much of the particulate load is deposited on the shallow continental shelf regions of the oceans.

TABLE 1. Some fluxes in the major sedimentary cycle

Material	Geosphere receiving material	Flux in 10^{14} g per year
Suspended river solids	Oceans	180
Dissolved river solids	Oceans	39
Glaciers	Oceans	30
Continental rock and soil particles	Atmosphere	1–5
Sea salt	Atmosphere	3
Volcanic debris	Stratosphere	0.036
Volcanic debris	Atmosphere	<1.5
Wastes of society (excluding fossil fuels)	Hydrosphere, lithosphere, atmosphere	30
Wastes dumped from ships	Oceans	1.4
Carbon and fly ash from fossil fuel combustion	Atmosphere	0.25
Industrial particulates total	Atmosphere	0.54
<2 μm		0.12

Adapted in part from E. D. Goldberg, 'Man's Role in the Major Sediment Cycle', in D. Dryssen and D. Jagner (eds.), *The Changing Chemistry of the Oceans*, p. 267–88, New York, N.Y., Wiley-Interscience, 1972.

Most of the world's rivers flow into the Atlantic Ocean (Fig. 2). (The Nile is included for historical purposes only since its flow has been drastically reduced by the construction of the Aswan Dam.) The Amazon river accounts for over 10 per cent of the total and it drains into the equatorial Atlantic, as does the second major river, the Congo. The Pacific Ocean receives only a modest amount of river-borne substances. In addition, most of these Pacific rivers enter marginal basins such as the Yellow Sea where much of their particulate load is trapped. Many other rivers enter marginal seas, bays or estuaries before entry to oceanic coastal waters. For example, the Rhine enters the North Sea; the Mississippi, the Gulf of Mexico. Such water bodies act as holding zones for pollutants (particularly dissolved phases), before admission to the coastal or open ocean. The residence time of the waters of the North Sea before mixing with the Atlantic Ocean is about three years. For Puget Sound, Washington, the replacement time of the waters is estimated to be about six months.

Many rivers of the world can no longer be considered to be in their natural states with regard to their dissolved and particulate loads. Northern-hemisphere rivers draining industrialized areas appear to have had their compositions substantially altered by man. Berner (1971) indicated that the sulphate fluxes, normalized to chloride fluxes, of North American and European rivers were significantly higher than those of African and South American rivers. For the former, there was an increase in sulphate levels of about 25 per cent (again compared to chloride) which was attributed to a source of fossil fuel combustion. Most probably these same rivers have increased particulate loads as a consequence of changes in land use, which have resulted in increased erosion (see below).

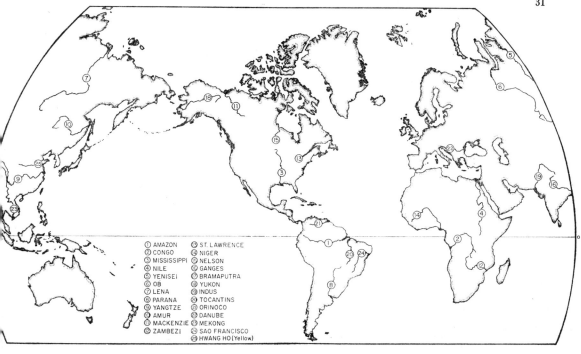

① AMAZON   ⑬ ST. LAWRENCE
② CONGO   ⑭ NIGER
③ MISSISSIPPI   ⑮ NELSON
④ NILE   ⑯ GANGES
⑤ YENISEI   ⑰ BRAMAPUTRA
⑥ OB   ⑱ YUKON
⑦ LENA   ⑲ INDUS
⑧ PARANA   ⑳ TOCANTINS
⑨ YANGTZE   ㉑ ORINOCO
⑩ AMUR   ㉒ DANUBE
⑪ MACKENZIE   ㉓ MEKONG
⑫ ZAMBEZI   ㉔ SAO FRANCISCO
   ㉕ HWANG HO (Yellow)

Fɪɢ. 2. The principal rivers of the world ranked in order of annual discharge.

WIND SYSTEMS

The principal wind systems tend to carry their materials along lines of latitude, although they do meander to the north and to the south from year to year, with a consequent spreading of their loads. The amount of time involved in wind transport of materials may be gained from the observation that radio-active debris introduced to the troposphere by a Chinese nuclear detonation at Lop Nor (40° N. and 90° E.) in May 1965 encircled the earth in about three weeks with an average velocity of 16 m/s (Cooper and Kuroda, 1966). Fallout of debris during its travels was observed in Tokyo (36° N. and 140° E.) and Fayetteville, Arkansas (36° N. and 94° W.).

The three principal wind systems are: The trades, which blow between 30° N. and 30° S. decreasing in intensity with increasing latitude. They blow from east to west. Reversals in directions can take place in the upper troposphere, except near the equator where their directions are nearly always easterly.

The mid-latitudinal westerlies, the jet streams, prevail between 30° N. and 70° N. and between 30° S. and 65° S., often with greater intensity in the upper troposphere than in the lower troposphere. Wind speeds of over 100 m/s have been observed, with speeds of 40 m/s common in their central parts.

The polar easterlies are near surface winds which occur at 70°–90° N. and 65°–90° S. They decrease in intensity with height and at about 3 kilometres they often reverse their directions.

In addition to these three systems there are the continental monsoons, such as those of India, whose flows are determined by continent-ocean temperature differences. They can reverse their direction between summer and winter.

Cyclonic and anticyclonic eddies occurring seasonally cause north–south movements. Generally, there is a slow drift of air toward the poles in the lower stratosphere between the equator and 30° north or south latitudes. Particles introduced into the lower stratosphere or upper troposphere near the equator can drift into both hemispheres. Particles introduced at about 30° in either hemisphere are generally restricted to a poleward direction in their longitudinal motions.

GLACIERS

Glaciers are responsible for the transport of materials they pick up in the polar regions to latitudes of about 50°. A greater amount of natural glacial transport of weathered materials takes place in the Antarctic regions. Glacial transport has not hitherto been a factor in pollutant studies.

OUTFALLS

The impact of domestic and industrial outfalls upon the coastal ocean can often be defined by the concentration 'halos' of organic matter, micro-organisms, heavy metals and other waste products in the waters and sediments adjacent to the outfalls. Particulate materials released from sewer outfalls have been observed as far as 10 kilometres from the discharge sites. The patterns of concentration reflect both discharge levels and local current systems.

SHIPS

Discharges from ships, deliberate as well as unintentional, are responsible for dispersion of materials both locally and globally. Most dumping from ships occurs over the continental shelf, frequently in estuarine regions. As mentioned in Chapter 1, these are the productive areas of the oceans. The annual disposal of the approximately 6 million tons of litter from ships is soiling coastal areas (see Chapter 7). The intentional dumping of materials in United States coastal waters quadrupled between 1949 and 1968 (CEQ, 1970). Dredge spoils accounted for about 80 per cent of the total, with industrial and domestic sewage wastes amounting to 10 per cent each.

The largest source of sediments entering the North Atlantic (excluding the Equatorial Atlantic) appears to be the dumping of wastes into the coastal ocean near New York City (Gross, 1970). About 8.6 million tons were disposed of each year between 1964 and 1968. The wastes were generated by a population of about 9 million people at an average rate of 2 kilograms per person per day. About 5 million tons of wet sewage is annually dumped into the outer Thames estuary (Shelton, 1973), perhaps the second largest source. One of the ubiquitous ingredients found in the sediments underlying the British dump sites were the husks of tomato seeds, highly resistant to degradation and therefore useful as a tracer of the dispersion of dumped materials.

Natural and pollutant fluxes

The amount of materials moved about the earth's surface by the activities of man is about one-tenth of that involved in the major weathering cycle (Table 1). The rivers carry, primarily to the continental shelf regions, around 20 billion tons of dissolved and solid substances annually. The winds carry between 0.1 and 0.5 billion tons of rock flour and soil materials per year over transoceanic or transcontinental distances. Glaciers transport around 3 billion tons per year of crustal material to the oceans, primarily in the Antarctic region. The discards of society (excluding the combustion products of fossil fuels) are about 3 billion tons per year, or more than 1 cubic kilometre.

A substantial amount of these discards eventually reaches the ocean system. Naturally occurring solids are dispersed to the atmosphere as a consequence of volcanic activity, of wind action upon exposed crustal materials, and of the injections of salt from the oceans. The fluxes of sea salt (0.3 million tons per year) and of rock flour and soil materials from the continents (0.1–0.5 million tons per year) are higher by about a factor of ten than man's emissions from industrial activity (0.012 million tons per year for small particles subject to long-range transport) and from fossil fuel combustion (0.025 million tons per year). These fine particles originate principally from the production of crushed stone, iron, steel and cement. Such particles are preferentially emitted from activities involving combustion, condensation and vaporization rather than from mechanical processes. The oxides of silicon, iron, aluminium, calcium and magnesium, and calcium carbonate, are the dominant materials.

The fluxes of crustal materials measured today may not be representative of natural processes inasmuch as they may have increased as a consequence of historical and contemporary changes in land use. The transformation of forest to croplands and grazing areas increases the erosion rate approximately tenfold. The construction of roads enhances erosion processes by one to two orders of magnitude. About one-quarter of the United States has undergone such transformation or has been given over to road building, so there may be an anthropogenic increase in the dispersion of surface materials to its winds and to its rivers. Similar situations undoubtedly exist in other parts of the world.

The formulation of mass-balance models

Mass balances provide a dynamic description of a pollutant's dispersion in the environment. They serve as a basis for the development of monitoring strategies. As we shall see, they can pinpoint areas where surveillance measurements must be made and assist in the determination of the frequency of such measurements.

The calculation of mass balances involves equating the inputs of natural or polluting materials from the continents to such reservoirs as the hydrosphere, atmosphere and biosphere with the fluxes out of the reservoirs and with their reservoir concentrations. Where only the rates of movement between the continents and these reservoirs and between the reservoirs themselves are considered, the mass-balance computation takes the form of a flow diagram. When the amounts and residence times in reservoirs are sought, the formulation is known as a mass-balance model. Mass balances can be constructed for both steady-state and transient situations.

Hitherto, most mass-balance calculations for pollutants have been based upon limited data. For only a very few substances have there been sufficient measurements to accept with confidence all of the details of the model. On the other hand, the obvious weakness of those models based upon scant data has in some instances stimulated concerned scientists to make additional measurements.

PRODUCTION AND USE DATA

The amount of a material inadvertently or deliberately released to the environment by man can sometimes be estimated from the production and use data. A simple case involves the gaseous chlorofluorocarbons used primarily as aerosol propellants. Since

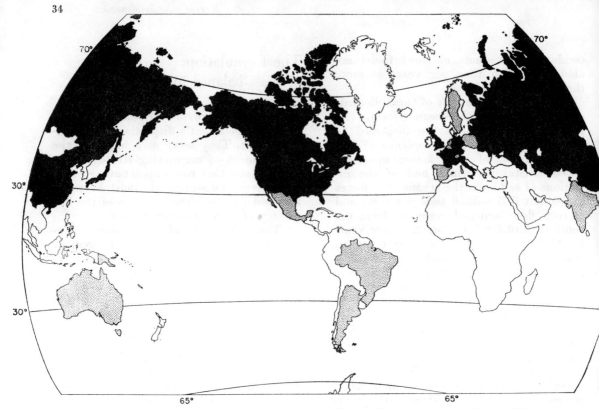

Fig. 3. Gross national product: solid enclosures indicate GNPs greater than $60 billion annually; light shading, between $19 and $60 billion; and unshaded, less than $19 billion annually.

virtually the total production is released to the atmosphere, a knowledge of the production figures gives the atmospheric flux.

On the other hand, a knowledge of the annual world production of petroleum does not lead to an estimate of losses to the environment, since the percentage lost has not been adequately determined. Direct measurements of losses such as those from ships as well as measurements of man-mobilized petroleum in various reservoirs, such as the amount of petroleum in sewer outfalls, has yielded an estimate of petroleum flows to the environment. Nevertheless, present and anticipated production and use data for petroleum will be valu-able in predicting future leakages, where estimates of present-day losses have been made and can be extrapolated to future situations.

In addition to the production and use statistics, the geographical sites of production and use are important in the formulation of mass-balance models. For example, nearly all energy production and material utilization by man takes place in the mid-latitudes of the northern hemisphere. This can be seen through an examination of the geographical distribution of the gross national product (GNP) of the world's nations (Fig. 3). The assumption is made that the energy production and material utilization are directly related to the GNP. Of the main industrial countries, only Australia, Brazil, Argentina and Mexico do not lie within a belt between latitudes 30º and 70º N. If we make the additional assumption that leakages to the environment of material wastes and combustion materials from energy production are proportional to the GNP, then a latitudinal pollution band has been defined. In formulating a global study of the atmospheric dispersion of a synthetic organic chemical today, emphasis should be placed on sample collection from the northern rather than the southern hemisphere, especially where steady-state conditions have not been attained.

Coolant water-discharge pipes from nuclear reactors sited on marine coasts are essentially point sources for the introduction of radio-active nuclides. Their locations indicate where to make measurements for inclusion in mass-balance calculations.

Production and use statistics (where the information is not proprietary) are available from various dispersed sources: For the United States, the Tariff Commission produces a bi-annual report on the production and sales of synthetic organic chemicals. The periodic *Minerals Yearbook* issued by the United States Department of the Interior has information on the uses of inorganic materials on both a national and world-wide basis. The Organization for Economic Co-operation and Development (OECD) periodical, *The Chemical Industry*, has limited data on world-wide production of some chemicals. Programmes directed by international organizations often have collated statistics on specific groups of substances: Food and Agriculture Organization (FAO) on pesticides; World Health Organization (WHO) on chemicals used in public health; International Atomic Energy Agency on radio-active materials; and Intergovernmental Maritime Consultative Organization (IMCO) on petroleum and hazardous substances involved in marine transportation.

RESERVOIRS

Following the accumulation of production and use data and the identification of the chemical and physical forms of the pollutant, the next step in the construction of mass balance or flow charts is the identification of the reservoir of the system into which the pollutant flows. For global problems, the oceans may be treated as an entity or divided up into compartments which might include deep and mixed layers, coastal and open-ocean waters, and combinations of various basins. For models involving estuaries, bays or the coastal zone, other groupings of reservoirs may be called for.

The atmosphere may be subdivided into the troposphere and stratosphere or treated as a single reservoir.

The biosphere may be compartmentalized into marine and terrestrial components. Certain collectives of the biosphere may accumulate large amounts of the pollutant and as a consequence be designated as a reservoir. For example, the photosynthesizing phytoplankton are a reservoir for the carbon dioxide produced in burning fossil fuel.

The solid phases of the earth, including land masses, the sediments, or even the continental ice sheets, may be utilized as reservoirs. Sedimentary and land domains may be subdivided into reservoirs where significant amounts of the pollutant might accumulate. For example, the sand and the clay deposits have widely varying surface areas per gramme of material—areas that can sorb widely varying amounts of a pollutant.

FLUXES

At steady state the amount of material entering a reservoir is compensated by the removal of an equivalent amount out of the reservoir or by its destruction within the reservoir. For purposes of approximation, this argument can also serve in non-steady-state situations over short time periods.

The flux of a pollutant between two reservoirs can often be estimated in several ways. Where there is mixing between water masses and the substance follows the water mass without fractionation (i.e. it is a conservative species), the time constants of the natural mixing process can be used. Other fluxes, such as evaporation from solution or washout from the atmosphere in rain, can be estimated from laboratory or field experiments.

First-order kinetics are assumed in most models, i.e. the amount transferring per unit time from one reservoir to another is directly proportional to the amount in the generating reservoir.

Several methods have been developed to estimate the transfer of materials from the atmosphere to the oceans by means of washout with rain, snow or sleet. These schematizations provide approximate values of atmospheric fluxes to marine waters where measurements of a substance's concentration in air or in precipitation are available from adjacent air masses. Most of the data that are fed into these computations have been obtained in the middle latitudes of the northern hemisphere. Thus, this application is more reasonable there than in areas south of the equator.

One method (Goldberg, 1972) assumes forty rainfalls per year, which sweep out all materials from a height of 5 kilometres. For a pollutant whose concentration in the atmosphere is C_i g/cm³, the atmospheric flux to the oceans through each square centimetre of surface is given by

$$F = C_i \cdot 2 \times 10^7 \text{ g/cm}^2 \text{ per year.}$$

For heavy metals this simple calculation has produced results that are in remarkable agreement with measurements of heavy metal fluxes in rainfalls (Lazrus *et al.*, 1970. Also see Table 2). Bruland *et al.* (1974) observed that the trace metal concentrations

TABLE 2. Trace metal concentrations in air and their environmental fluxes

| Metal | Aerosol concentrations in ng/m³ | | | Washout fluxes g/cm² per year | |
	Wraymires, England (1)	San Francisco, California (2)	Reasonable value from (1) and (2)	Calculated	Santa Catalina, California
Pb	112		100	2	1.3
Cr	3.1	8.2	5	0.1	
Zn	103	136	100	2	4.6
Cu	34	50	50	1	0.5
Ni	7		7	0.14	0.24
Mn	14	17	15	0.3	0.36
Co	0.4	1	0.7	0.014	
V	10	5.4	7	0.14	
Al	335	863	600	12	
Fe	297	1,670	1,000	20	

Data from Wraymires taken from D. H. Peirson, P. A. Cawse, L. Salmon and R. S. Cambray, 'Trace Elements in the Atmosphere Environment', *Nature*, Vol. 241, 1973, p. 252–6; for San Francisco: W. John, R. Kaifer, R. Rahn and J. J. Wesolowski, Trace Element Concentrations in Aerosols from San Francisco Bay Area, *Atmospheric Environment*, Vol. 7, 1973, p. 107–18; for Santa Catalina: A. L. Lazrus, E. Lorange and J. P. Lodge Jr, 'Leads and other Ions in United States Precipitation', *Envir. Sci. Technol.*, Vol. 4, 1970, p. 55–8.

in the atmosphere over the British Isles and California are similar. Further, recent work on Antarctic atmospheric dust samples reveals a close correspondence in their heavy-metal concentrations with those of the British Isles and California. This leads to the interesting possibility of a world-wide (pollutant plus natural) dust burden in the world's air. Feeding a reasonable aerosol burden value from the British and California data into the above equation yielded a washout flux similar to that observed by Lazrus on Santa Catalina Island during a six-month period in 1966 and 1967 when the solids, greater than 5 microns, were removed prior to analysis.

A second method employs the concentrations of the material in rain. Assuming an average annual rainfall of 75 centimetres the flux of materials to a square centimetre of surface is given. $F = 75\ C_2\ \mathrm{g/cm^2}$ per year, where C_2 is the pollutant concentration in rainfall in $\mathrm{g/cm^3}$ of rain. Both models give similar results for the fluxes of mercury to the earth's surface (Weiss *et al.*, 1971).

CHEMICAL AND PHYSICAL FORMS

The pollutant's chemical and/or physical states during and subsequent to release to the environment can play an important role in mass-balance calculations. Particle size can determine the distance of transport whether by air or by water. This is illustrated in the flow diagram of lead (see below on models) where both large and small particles are exhausted from automobiles to the atmosphere. The larger particles fall back to the earth's surface within a short distance from the automobile; the smaller particles are carried by wind systems to great distances from the road. A further complication involves the emission of volatile lead species whose environmental chemistries are quite different from those of the particles.

Environmental behaviours of some pollutants have played havoc with the intuition of scientists. At normal temperatures and pressures DDT is a solid with an extremely low vapour pressure (about 10^{-7} millimetres of mercury; see Chapter 3). Yet, its movement about the surface of the earth is primarily in the vapour state. The initial measurements of its atmospheric concentrations were low by a factor of 1,000, since only the particulate forms of DDT were analysed.

The pollutant forms of an element may not be the same as those naturally occurring. The chemical forms of artificially produced radio-active zinc in sea water differ substantially from those of the naturally existing stable isotopes. The production of methylated forms of mercury in man's wastes and by micro-organisms in natural situations came as a surprise to investigating scientists, normally used to considering only inorganic mercury compounds in our surroundings (see Chapter 5).

The form initially emitted to the environment can alter during transport or during residence in a reservoir. Radio-active nuclides can decay to daughters whose chemistries differ radically from those of the parent. Sulphur released during the burning of coal, oil and natural gases enters the atmosphere as sulphur dioxide. There it is quickly oxidized to the gas sulphur trioxide which then takes up water to form sulphuric acid. This latter substance takes the form of an aerosol, a particulate species in the atmosphere. The acid enters the ocean system either by dry fallout or by washout with precipitation, and quickly dissociates into sulphate and hydrogen ions.

BIOACCUMULATION

Marine organisms can accumulate chemical species in amounts far exceeding their sea-water concentrations. The term 'concentration factor' has been used to quantify this phenomenon and may be defined as the concentration of the chemical species in the organism or one of its components divided by the sea-water concentration in the environment from which the organism was taken. The organisms may take up the chemical directly from sea water or from their food. Thus, depending upon the composition of the food, concentration factors can vary from one environment to another for a specific organism with a given chemical.

A noteworthy characteristic of the bio-accumulation process is the remarkable specificities for a given chemical shown by organisms. For example, it has been known for about a half a century that some species of tunicates take up vanadium from sea water where its concentration is submicromolar. Other species of tunicates concentrate vanadium's periodic table neighbour niobium, whose sea water content is one-tenth less. Still others can accumulate neither element. No tunicate has yet been observed that can dramatically amass both. Some species of oysters are enriched in zinc, some sea-weeds in ruthenium, some sea grasses in beryllium. DDT is amassed from sea water by the higher gilled organisms.

The processes involved in this accumulation and in the specificity displayed by individual species are poorly understood. The selectivity problem may involve macromolecules which provide both an appropriate geometry and the necessary chemical linkages for the uptake of a specific chemical species from sea water. For example, the ability of the blood pigment haemocyanin to accommodate copper is presumably a function of the geometry of the macromolecular protein since haemocyanin itself contains no low molecular weight complexing groups. On the other hand, the enrichment of heavy metal ions by marine organisms in general reflects their ion exchange properties with solids as observed in the laboratory. The accumulation processes probably initiate on the external

surfaces of organisms where exposed chemical functional groups can chelate or complex dissolved species from sea water. Such surfaces include the mucus layers of animals and the polysaccharide sheets coating marine plants. A second form of assimilation of species from sea water involves the intake of particles, such as the micron- and submicron-sized iron hydroxide flocs and clays, and the materials sorbed onto them. Filter-feeding organisms, such as clams and oysters, have the ability to accumulate the particles and the chemicals associated with them.

Three mass-balance models

Three mass-balance models are presented to illustrate the various types of information required for their formulation. Their descriptive and predictive capabilities will also be considered. The first defines the flow of polychlorobiphenyls (PCBs) in and about the environment where use and disposal information has been coupled to production figures. The second considers the search for reservoirs for carbon dioxide produced through the consumption of fossil fuels. Thirdly, we shall examine a flow sheet for lead in a coastal system, emphasizing the selection of reservoirs and the identification of the chemical forms involved.

THE PCB MODEL

The principal routes of the PCBs from the United States to the oceans were identified by Nisbet and Sarofim (1972) and incorporated in their model (Fig. 4). The important entries took place through rivers, domestic or industrial sewage outfalls and the atmosphere, as is illustrated in their routing sheet (Fig. 5). The variety of PCB uses, coupled with different environmental paths for each, emphasizes the difficulties in arriving even at order-of-magnitude estimates

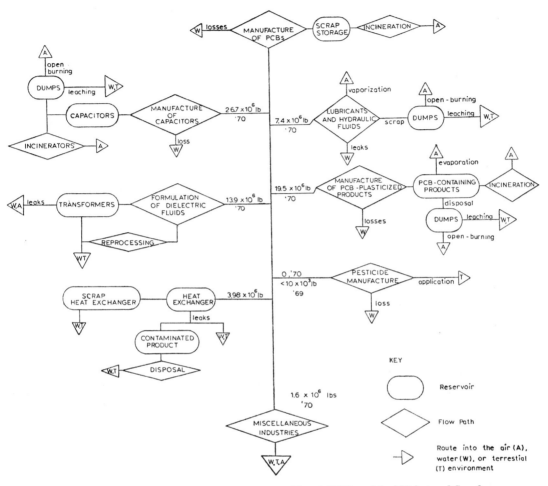

FIG. 4. PCB model of Nisbet and Sarofim.

of fluxes. The following assumptions were made. Transformers utilize PCBs as dielectrics in what may be considered relatively closed systems. Nisbet and Sarofim estimate that 10 per cent of the production of PCBs used as transformer oils replaces the oil that escapes or breaks down. The remaining 90 per cent of production enters new units. They estimate the useful life of a capacitor containing PCBs, such as those used in fluorescent lighting fixtures,

to be under a decade. On the basis of a doubling of sales every ten years, they estimate that the rate at which capacitors are discarded (primarily into land-fill dumps) is equal to half the rate of production. The rate of vaporization of a PCB plasticizer is taken to be 10–20 per cent of the annual production. Environmental losses of hydraulic fluids and lubricants were taken as equal to their production.

Using these and other rough estimates,

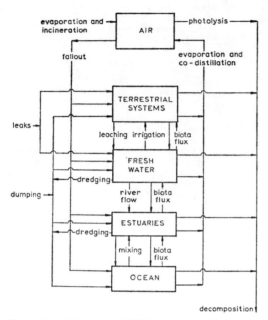

FIG. 5. Possible routes of PCBs into the environment as illustrated by Nisbet and Sarofim.

only about 20 per cent of the 1970 gross production appears to have been added to the PCBs in use; the remainder is assumed to have been discharged to the environment. The fluxes were estimated to be:

$1-2 \times 10^3$ tons per year through the evaporization of plasticizers.

$4-5 \times 10^3$ tons per year by leaks and disposal of hydraulic fluids and lubricants.

22×10^3 tons per year by disposals in incinerators, dumps and sanitary landfills.

Of this last category, it was estimated that 10–20 per cent ($2-4 \times 10^3$ tons per year) were destroyed by burning and 2 per cent (4×10^2 tons per year) were vaporized, mainly by open burning of wire scrap, auto components and other PCB-containing materials in dumps.

These data were then converted into environmental fluxes as follows:

$1.5-2 \times 10^3$ tons per year into the atmosphere.

$4-5 \times 10^3$ tons per year into fresh and coastal waters.

18×10^3 tons per year into dumps and landfills.

The next step in the model formulation is the estimation of fluxes within the environment and the levels in the reservoirs. The PCBs have chemical and physical properties closely resembling those of DDT and its degradation products. This allowed estimation of some rates of transport and the routes from corresponding information on DDT. On this basis, the annual fluxes to the reservoirs were computed to be:

United States soils: 1.5×10^4 tons.

Oceans adjacent to North America: 1.5×10^4 tons.

United States fresh water: 10^2 tons.

United States fresh-water sediment: 2×10^4 tons.

United States biosphere: $<10^3$ tons.

On the basis of environmental degradation studies, about one-third of the PCBs entering the atmosphere and about one-half entering natural waters have disappeared. Using an analogy with DDT, Nisbet and Sarofim estimate that one-quarter of the total amount of PCBs introduced to the air and not destroyed there (5×10^3 tons) have entered the oceans.

The major portion of the 1.5×10^4 tons entering the oceans from all sources probably ends up primarily in the North Atlantic Ocean. Because of the prevailing westerly winds over the United States and of the industrial activities of the eastern United States coastal communities, this amount may be compared with that put forth by Harvey *et al.* (1973) who estimated on the basis of field measurements that there are 2×10^4 tons of PCBs in the upper 200 metres of water in the North Atlantic. Taking a residence time of approximately a year in the surface waters, this is a reasonable agreement. The bulk of the PCBs introduced to the open-ocean water would then be in deeper waters.

THE CARBON DIOXIDE MODEL

The combustion of fossil fuels (coal, oil and natural gases) is introducing carbon dioxide into the atmosphere at such a rate that present environmental concentrations of around 320 p.p.m. CO_2 are increasing by 0.7 p.p.m., or 0.2 per cent, per year. Possibly this added burden of CO_2 may eventually have an effect upon climate by the increased absorption of radiation with a subsequent heating of the upper atmosphere. Three primary reservoirs for the carbon dioxide have been identified: the atmosphere, the biosphere and the oceans. Several models have been devised to describe the flow and content of carbon dioxide in these reservoirs, primarily with the purpose of predicting future levels. The model of Machta (1972) will be considered.

The applicability of this model to the prediction of future atmospheric concentrations of carbon dioxide depends upon a knowledge of (a) the input of carbon dioxide from the burning process; (b) the fluxes of carbon dioxide between each reservoir; and (c) the time stability of each reservoir.

The model is illustrated in Figure 6. First-order kinetics are assumed between the troposphere and mixed layer of the oceans, between the mixed and deep layers of the ocean, and between the troposphere and stratosphere. The reaction rate constants (the fraction of the carbon dioxide in one reservoir that is transferred to another reservoir in a unit of time) were derived from meteorological and oceanographic studies. A novel part of this model is the definition of the biosphere as a reservoir to which a mass of carbon is transferred each year equal to the primary production. The carbon can be transferred from the troposphere to the terrestrial biosphere or from the mixed layer of the ocean to the marine biosphere.

The partition of carbon dioxide between the various reservoirs has been estimated

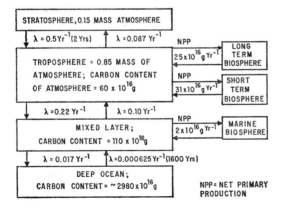

FIG. 6. Carbon dioxide model of Machta with some later modifications. The rate constants are first order.

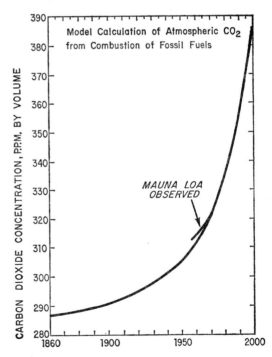

FIG. 7. The predicted and observed carbon dioxide changes in the atmosphere as a function of time (Machta, 1972).

for 1970: 55 per cent of the gas produced in fossil-fuel combustion up to this date was in the atmosphere; 15 per cent in the biosphere; and 30 per cent in the oceans (25 per cent in the mixed layer and 5 per cent in the deep layer).

The Machta model has been used to predict the atmospheric levels of CO_2 up to the year 2000 assuming a 4 per cent annual increase in the combustion of fossil fuels up to 1979 and 3.5 per cent between 1980 and 1999 (Fig. 7). The CO_2 levels of prediction and observation were normalized to each other in 1958 with an air concentration of 313 p.p.m. at the Mauna Loa Observatory.

A LEAD MODEL

The flow of lead through the Los Angeles area to the near-by coastal and open ocean has been modeled by Huntzicker *et al.* (1975). Their assessment of the impact of

the lead technology on the coastal zone provides a pattern applicable to other pollutants. Their model is schematized in Figure 8.

Pollutant lead is assumed to result entirely from the combustion of lead alkyls in automobiles. In the Los Angeles area 23.6 ± 2.4 tons were consumed per day in 1972. Of this, 17.9 tons per day were emitted from automobile exhausts to the atmosphere (17.3 tons per day as particles and the remainder as vapours, presumably as alkyl leads); 5.7 tons per day were accumulated in automobiles (in the oils, motors, exhaust pipes, etc.). The eventual fate of this lead has not yet been resolved.

The aerosols exhausted from automobiles have the following size distribution: >9 μm, 57 per cent; 0.3–9 μm, 27 per cent; <0.3 μm, 16 per cent.

However, measurements in air samples collected in the Los Angeles area showed only 2 per cent of the lead particles greater than 9 μm. Consequently, the greater than 9 μm fraction (9.9 tons per day) was assumed to have been deposited near the road (near or road fallout). Of this amount, about 8.4 tons per day fell directly upon the road and 1.5 tons per day on the land adjoining the road. The lead that fell on the streets may have been partially removed by street-cleaning operations or by runoff of surface waters. These latter numbers were determined by measurements of lead fallout as a function of distance from a Los Angeles area motorway.

'Far fallout' is defined as the lead that deposits at locations distant from the source. Measurements of lead fallout on land in the Los Angeles area resulted in the estimate of 2.0 ± 0.9 tons per day over a total land area of 4,680 square kilometres for the far fallout rate.

The rate of removal of lead from the area was determined by using carbon monoxide as a tracer. Airborne lead and carbon monoxide, which like lead derives en-

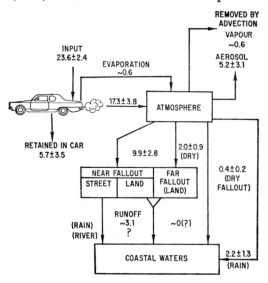

FIG. 8. The flow of automotive lead through the Los Angeles area. Numbers are in tons per day (Huntzicker *et al.*, 1975). Reprinted with full permission. Copyright by the American Chemical Society.

tirely from automobile emissions, co-vary in their atmospheric concentrations. On this basis Huntzicker *et al.* estimated that 5.2 ±3.1 tons per day of particulate lead were removed by advection. The vaporized lead alkyls from the exhausts, which are presumed to be essentially inert in the atmosphere, were also removed in this way.

In the model it was postulated that lead entered the coastal waters by at least five routes: dry atmospheric fallout; direct washout from the atmosphere with precipitation; storm-water runoff; river runoff and municipal sewage discharges. By direct measurement of dry fallout on islands in the coastal region and by extrapolation of land values, fluxes of 0.4±0.2 tons per day were found. Similarly, measurements of lead in rainfall, made in the same way as the dry fallout determinations, gave a rain removal to the coastal zone of 2.2±1.3 tons per day.

Lead assays in storm runoff waters from the Los Angeles area indicated an upper limit or maximum daily input to the coastal waters of 3.1 tons per day. Although the river runoff of lead to the coastal waters is not known, an outfall flux of 0.6 tons per day is found on the basis of measurement of lead in sewage discharges.

Thus, the total amount of lead entering the coastal waters off Los Angeles is about 5.7 tons from a daily combustion of nearly 24 tons. About 10 per cent of this, or 0.5 tons per day, has been estimated to be accumulating in the coastal sediments. Hence, about 90 per cent of the lead introduced to the oceans off Los Angeles either remains within the coastal waters or is eventually introduced to the open ocean.

The historical record

Sedimentary columns often contain continuous, chronological records of pollutant concentrations. The coastal deposits beneath waters of high biological activity are especially attractive for marine studies. In cases where the sediment pore waters become anoxic (i.e. devoid of dissolved oxygen gas and hence incapable of supporting aerobic animal life), the consequent absence of animal burrowing activity may result in an undisturbed depositional record. Such deposits are located in areas of restricted water circulation, a situation that does not permit renewal of oxygen in the waters overlying the sediments through their contact with the atmosphere or through their mixing with oxygenated waters. Examples of such deposits include most fjords, the inner basins of southern California, and some shelf sediments under highly productive waters. Where the pollutants have short residence times in the waters, their fluxes to coastal deposits are readily measurable. For these pollutants, records of past oceanic concentrations can be reconstructed when present-day ocean-water and surface-sediment (i.e. recently deposited) concentrations are known.

Changes in lead fluxes to the environment from the combustion of lead alkyls in gasolines have been measured in anoxic basin sediments west of Los Angeles, a densely populated urban community (Chow *et al.*, 1973). These deposits accumulate at rates of millimetres per year, and annual strata can be assigned an age by counting annual layers or by radiometric dating techniques in the absence of annual layers. Samples that have deposited within a given year can often be obtained.

The changes in lead concentrations as a function of depth are shown in Figure 9 for three southern California basin deposits as well as for one off Baja California, Mexico, where the adjacent lands are only sparsely populated. The rates of lead accumulation began to increase in the 1940s for the southern California sediments, an effect not evident in the Baja California deposits used as a control. Fluxes of lead can be ascertained by subtracting the pre-

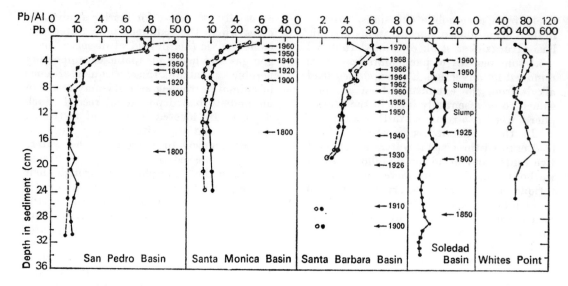

FIG. 9. Lead concentrations in sediments from Santa Monica, San Pedro and Santa Barbara basins off southern California, from the Soledad Basin off the coast of Baja California, and from a site near the Whites Point, Los Angeles County, Outfall. The Soledad Basin may be considered a non-polluted area, and it is not adjacent to continental areas of intense industrial activity as are the southern California basins. (Reproduced from Chow *et al.*, 1973. Copyright 1973 by the American Association for the Advancement of Science.)

pollution concentrations of lead, say those in the strata deposited before 1940, from the anthropogenic concentrations deposited after 1940, using the sedimentation rates. Of the 24 tons of lead burned daily in the Los Angeles area, 0.5 tons are calculated to have been deposited in the sediments of the 12,000 square kilometres coastal zone off Los Angeles. The bulk of the lead is accumulated in other reservoirs. (See above on models.)

Many coastal areas do not possess undisturbed, continuously accumulating sediments and the historical record of their pollutant injections must be sought elsewhere, often with limitations. For example, permanent snow fields (glaciers) contain records of atmospheric fallout (Windom, 1969). These deposits occur at all latitudes, a useful attribute. The strata can be as-signed ages by radiometric dating techniques, by counting annual layers, where visible, or by oxygen isotope analysis. Depositional rates of up to tens of centimetres per year often allow seasonal strata to be defined. It is reasonable to assume that ocean waters adjacent to permanent snow fields will receive equivalent amounts of atmospheric fallout.

The records of anthropogenic lead fall-out from the atmosphere are preserved in the Greenland glacier (Murozumi *et al.*, 1969). Increasing entries to the atmosphere over the past several thousand years resulted from the increasing uses of this metal by man. The sharply increased lead flux attributed primarily to the combustion of lead alkyls in petrol, becomes evident after the 1940s (Fig. 10). Clearly, the ocean environment at the same latitudes

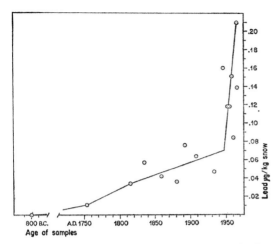

FIG. 10. The lead concentrations in dated levels of a Greenland glacier (Murozumi *et al.*, 1969).

near Greenland must have received similar amounts of lead fallout.

Pollution records in marine organisms have been sought through museum specimens. Both skeletal materials and preserved whole specimens have been used with varying degrees of success. The annual layers of corals deposited in Entiwetok Atoll have logged their exposures to radio-active fallout from bomb detonations through their uptake of strontium-90 (Knutson *et al.*, 1972). Bertine and Goldberg (1972) measured the concentrations of a group of heavy metals in species of the mussel *Mytilus* collected over the past hundred years in coastal European waters but were unable to discern any significant changes that might have been attributed to pollution. In principle, skeletal hard parts are preferable to whole plants or animals since

the latter are usually maintained in pickling solutions which may have contained quantities of the pollutant under study or which may have leached out the pollutant. However, the exoskeletal materials may record a different type of information with respect to the composition of their surroundings from that which the flesh may record. Each may be more or less useful for a specific contaminant.

The concern over man-mobilized mercury to marine waters stimulated a number of investigations of the pollution record held by deep-sea fish. The results showed that open-ocean fish collected over the past hundred years had consistent mercury concentrations within experimental error, suggesting that the mercury content of the open ocean has not been measurably affected by man's activities (see Chapter 5).

Whether meaningful results can be obtained from museum specimens is a question that has been raised by Gibbs *et al.* (1974). Specimens of lantern fish were maintained for a month in various typical preservatives (formalin, ethyl alcohol and isopropyl alcohol). Such preservation resulted in higher concentrations of cadmium, copper and zinc, lower concentrations of mercury, and more variable concentrations of lead than those in a set of unpreserved frozen specimens. They also analysed a set of museum specimens, which had been preserved for periods ranging between 2 and 85 years. Higher concentrations of the metals were generally found in older specimens. The authors suggested that not only the preservatives, but also metal tags and other artifacts in the containers of preserved fish, could alter the original composition of the fish.

Bibliography

BERNER, R. A. 1971. Worldwide sulfur pollution of rivers. *J. Geophys. Res.*, vol. 76, p. 6597–600.

BERTINE, K. K.; GOLDBERG, E. D. 1972. Trace elements in clams, mussels and shrimp. *Limnology and Ocean-ography*, vol. 17, p. 877–84.

BRULAND, K. W.; BERTINE, K.; KOIDE, M.; GOLDBERG, E. D. 1974. History of metal pollution in southern California coastal zone. *Environ. Sci. Technol.*, vol. 8, p. 425–31.

CEQ. 1970. *Ocean dumping, a national policy.* United States Council on Environmental Quality, October 1970.

CHOW, T. J.; BRULAND, K. W.; BERTINE, K.; SOUTAR, A.; GOLDBERG, E. D. 1973. Lead pollution: records in southern California coastal sediments. *Science*, vol. 181, p. 551–2.

COOPER, W. W.; KURODA, P. K. 1966. Global circulation of nuclear debris from the May 14 1965 nuclear explosion. *J. Geophys. Res.*, vol. 71, p. 5471–3.

GIBBS, R. H.; JAROSEWICH, E.; WINDOM, H. L. 1974. Heavy metal concentrations in museum fish specimens. Effects of preservatives and time. *Science*, vol. 184, p. 475–7.

GOLDBERG, E. D. 1972. Man's role in the major sediment cycle. In: D. DRYSSEN and D. JAGNER (eds.), *The changing chemistry of the oceans*, p. 267–88. New York, N.Y., Wiley-Interscience.

GROSS, M. G. 1970. *New York City—a major source of marine sediment.* Stony Brook, N.Y., Marine Sciences Research Center, State University of New York City. (Technical Report No. 2.)

HARVEY, G. R.; STEINHAUER, W. F.; TEAL, J. M. 1973. Polychlorobiphenyls in North Atlantic Ocean water. *Science*, vol. 180, p. 643–4.

HUNTZICKER, J. J.; FRIEDLANDER, S. K.; DAVIDSON, C. I. 1975. Material balance for automobile-emitted lead in the Los Angeles Basin. *Environ. Sci. Technol.*, vol. 9, p. 448–57.

JOHN, W.; KAIFER, R.; RAHN, R.; WESOLOWSKI, J. J. 1973. Trace element concentrations in aerosols from the San Francisco Bay area. *Atmospheric Environment*, vol. 7, p. 107–18.

KNUTSON, D. W.; BUDDEMEIER, R. W.; SMITH, S. V. 1972. Coral chronometers: seasonal growth banks in reef corals. *Science*, vol. 177, p. 270–2.

LAZRUS, A. L.; LORANGE, E.; LODGE Jr, J. P. 1970. Lead and other ions in United States precipitation. *Environ. Sci. Technol.*, vol. 4, p. 55–8.

MACHTA, L. 1972. The role of the oceans and biosphere in the carbon dioxide cycle. In: D. DRYSSEN and D. JAGNER (eds.), *The changing chemistry of the oceans*, p. 121–45. (Modifications to text in personal communication, 1974.)

MUROZUMI, M. T.; CHOW, J.; PATTERSON, C. C. 1969. Chemical concentrations of pollutant lead aerosols, terrestrial dusts and sea salts in Greenland and Antarctic snow strata. *Geochim. Cosmochim. Acta*, vol. 33, p. 1247–94.

NISBET, I. C. T.; SAROFIM, A. F. 1972. Rates and routes of transport of PCBs in the environment. *Environmental Health Perspectives*, vol. 1, p. 21–38.

PEIRSON, D. H.; CAWSE, P. A.; SALMON, L.; CAMBRAY, R. S. 1973. Trace elements in the atmosphere environment. *Nature*, vol. 241, p. 252–6.

SHELTON, R. G. J. 1973. Some effects of dumped, solid wastes on marine life and fisheries. In: E. D. GOLDBERG (ed.), *North sea science*, Cambridge, Mass., MIT Press.

WEISS, H. V.; KOIDE, M.; GOLDBERG, E. D. 1971. Mercury in a Greenland ice sheet: evidence of recent input by man. *Science*, vol. 174, p. 692–4.

WINDOM, H. 1969. Atmospheric dust records in permanent snowfields: implications to marine sedimentation. *Bull. Geol. Soc. Amer.*, vol. 80, p. 761–82.

3.

Halogenated hydrocarbons

Investigations of the environmental behaviours of the synthetic halogenated hydrocarbons have not only provided a series of scientific surprises, but they have also emphasized the need for production and use data to help us interpret adequately the existing information.

Perhaps the most unexpected result was the discovery that the principal mode of transport of DDT and its metabolites and the PCBs from the continents to the oceans appears to be in the vapour phase. The vapour pressures of these substances are unusually low: 1.5×10^{-7} mm of mercury for DDT and 10^{-4} to 10^{-6} mm of mercury for the PCBs. Such low vapour pressures initially appeared irreconcilable with the widespread occurrences of these synthetic compounds in marine organisms, from plankton in every ocean to the arctic seals and the antarctic penguins. Early workers assumed that DDT-type compounds were associated with solid phases in their analysis of atmospheric dusts and rains. More recent studies, however, have shown that nearly all of the DDT is in the gaseous phase in the atmosphere and that its vapour pressure can account for the observed atmospheric DDT burdens.

Also surprising was that discovery that the heavier halogenated hydrocarbons like DDT and the PCBs are not necessarily concentrated increasingly in organisms as one goes higher up in the food chain. To the contrary, it appears that there is some sort of equilibrium established between the sea water and some organisms such as gilled fish for these compounds, possibly similar to their distribution between water and oil.

The difficulties in obtaining world-wide production and use data for these compounds have restricted our ability to understand their distribution in the marine system. Since the amounts of DDT and PCBs produced in the world are not known, accurate mass-balance calculations have not been possible. This problem has been further aggravated by the insufficient number of analyses made in sea water, primarily due to the demanding nature of the analytical procedures which few chemists have been able to master.

Both difficulties can be resolved in principle. The analytical techniques are now being mastered by a group of laboratories. The production and use data are being estimated by such international organizations as FAO and WHO with what appears to be reasonable accuracies.

Among the heavier halogenated hydrocarbons there are others used extensively as biocides (dieldrin, toxaphene, chlordane, endosulfan, hexachlorobenzene and mirex) or as industrial chemicals such as the chlorinated paraffins. However, we do not find them generally in environmental samples. We shall consider two as examples of special sets of circumstances. Hexachlorobenzene is somewhat widely distributed but is of interest inasmuch as it enters the environment most probably as a waste product from the production of other chemicals or as an impurity in manufactured products. The dispersion of mirex about the marine system follows entry on a regional basis, either as a consequence of production or of use.

Another set of synthetic halogenated organic compounds is creating some concern—the low molecular weight halocarbons, including the fluorinated species used as aerosol propellants, dry-cleaning fluids and industrial solvents. Their ubiquitous distributions in surface waters and in the marine and terrestrial atmospheres coupled with residence times of the order of years in the air, indicates that serious consideration should be given to their possible effects upon living systems. The atmospheric contents of one group, the chlorofluorocarbons, have been implicated in decreasing the ozone level of the stratosphere.

DDT and PCBs

DDT (pp′ DDT, 2, 2-bis (p-chlorophenyl)-1, 1, 1-trichloroethane) is degraded in the environment by solar radiation and by the metabolic activities of organisms to DDE (1, 1-di-chloro-2, 2-bis (p-chlorophenyl) ethylene) and DDD (1, 1-dichloro-2, 2-bis (p-chlorophenyl) ethane) (see Fig. 11 below). The combination of DDT and degradation products is often referred to as DDT residues or total DDT (t-DDT) or is given similar appellations. Here we shall refer to them as DDT residues.

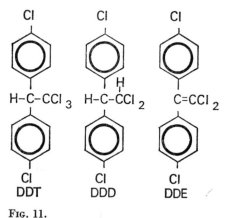

Fig. 11.

DDT USES

The two primary uses of DDT that have resulted in its environmental dispersal are in public health and agriculture. In the former it has largely been employed as a biocide against adult mosquitoes in malarial control. In agriculture one of its principal applications is as a biocide in the protection of cotton crops. Restrictions upon its agricultural uses have already been enacted in the United States, Canada, Japan, Sweden and the Soviet Union following observations of unwanted mortalities and morbidities among non-target organisms.

DDT PRODUCTION

The United States has been one of the principal sources of DDT (Table 3). Much was used elsewhere. Of the 63,400 tons manufactured in 1968, 49,600 were exported. These data were assessed by a United States group in 1970 (NAS, 1971)

TABLE 3. DDT production in the United States, 1944–68 (in thousands of tons)

Year	Production	Year	Production
1968	63.4	1955	59.0
1967	47.0	1954	44.2
1966	64.2	1953	38.4
1965	64.0	1952	45.4
1964	56.2	1951	48.2
1963	81.3	1950	35.5
1962	75.9	1949	17.2
1961	77.9	1948	9.2
1960	74.6	1947	22.5
1959	71.2	1946	20.7
1958	66.0	1945	15.1
1957	56.6	1944	4.4
1956	62.6	TOTAL	1,220

Adapted from NAS, *Chlorinated Hydrocarbons in the Marine Environment*, United States National Academy of Science, 438 p.

who concluded that at that time the total world production was probably no more than one and a half times that of the United States. As a first approximation, an integrated world production (i.e. the total amount of DDT since production began) was estimated to be 2.0×10^{12} grammes.

The United States export of 37,300 tons in 1969 was shipped as mass-produced technical-grade material and in formulations containing less than 75 per cent active ingredients, primarily for agricultural purposes, and in formulations containing more than 75 per cent active ingredients for malaria control purposes. Whittemore (personal communication, 1973) has proposed

the following breakdown, in tons technical grade equivalent, for 1969: technical grade, 12,700; formulations <75 per cent, 2,500; formulations >75 per cent, 22,100; giving a total of 37,300 tons.

Ten per cent of the technical-grade material and all of the formulations of more than 75 per cent active ingredients were ultimately used for public-health purposes. Thus, 13,930 tons were exported for other applications, primarily in agriculture.

About equal amounts of the DDT exported by the United States in 1969 for agricultural applications were utilized north and south of the equator as is shown by the following figures (in tons): North America, 4,400; South America, 4,700; Asia, 2,000; Africa, 2,600.

To arrive at an estimate of the current agricultural use of DDT, Whittemore has used as a basis its application in cotton protection. DDT has played a very important and frequently irreplaceable role in cotton agriculture. Brazil, for example, indicated that 70–75 per cent of its DDT was used on cotton crops. The following assumptions were used in Whittemore's calculations:

DDT applications to cotton are at the annual rate of 10 kg/hectare in Central America, 5 kg/hectare in South America and 0.5 kg/hectare in Asia and Africa.

Of DDT 75 per cent is used on cotton and the remainder on other crops in the major cotton-producing countries.

The DDT application rate for other crops in non-cotton-producing countries is assumed to be the same as that for other crops in cotton-producing countries.

Current-use estimates are given in Table 4 for Latin America (Central and South America), Africa and Asia.

Future estimates of DDT use depend upon projected cotton production. Whittemore, using FAO data, indicates an annual rate of growth in world cotton production between 1970 and 1980 of 1.6 per cent of which there is a 2.4 per cent in-

TABLE 4. Estimate of annual utilization of DDT in agricultural activities (in tons per year)

Region	DDT usage			
	Cotton-producing countries		Non-cotton-producing countries, other crops	Total
	Cotton	Other crops		
Central America	7,580	2,550	383	10,513
South America	18,800	6,200	1,180	26,180
Africa	2,186	729	605	3,520
Asia	5,568	1,523	410	7,501
TOTAL	34,134	11,002	2,578	47,714

Source: F. W. Whittemore (personal communication), Senior Officer, Plant Production Service, FAO, Rome (Italy).

TABLE 5. Estimate of future annual requirements for DDT in agricultural activities prepared in 1972 (in tons per year)

Region	DDT usage			
	Cotton-producing countries		Non-cotton-producing countries, other crops	Total
	Cotton	Other crops		
Latin America (Central and South America)	13,560	1,510	270	15,340
Africa	13,100	1,410	1,170	15,680
Asia	33,500	3,350	905	37,755
TOTAL	60,160	6,270	2,345	68,775

Source: F. W. Whittemore (personal communication), Senior Officer, Plant Production Service, FAO, Rome (Italy).

crease in the developing countries. The following assumptions guided his computations: (a) no economically acceptable substitute for DDT will become available in the foreseeable future; (b) the annual amount of DDT required for the protection of cotton is 3 kg/hectare; (c) the proportion of DDT used in agriculture for other crops will decrease from 25 per cent to 10 per cent; (d) DDT will be used at the same rate on all other crops in cotton-producing and non-cotton-producing countries.

The forecast indicates an over-all increase in agricultural DDT usage over the next decade, with a marked decrease in requirements for Latin America and an even more marked increase for Asia and Africa (Table 5).

The requirements for DDT as a residual spray against adult mosquitoes in antimalarial programmes for the decade 1971–81 has been predicted to be 470,000 tons (Rafatjah and Stiles, 1972). The use in 1970 was 42,600 tons, practically all as 75 per cent water-dispersible powder whose ability to be easily suspended and whose stability during packing, shipping and storage make it desirable. The estimated requirements are given in Table 6 for six regions of the world: Africa, America, South-East Asia, Europe, Eastern Mediterranean and Western Pacific. The annual needs for the 75 per cent water-dispersible powder are expected to increase towards a maximum in 1977 of 52,000 tons total. After 1977 a decrease is expected and in 1981, the last year of this forecast, the requirement will probably be about 29,000 tons.

The total predicted demands by country over this ten-year period are plotted in Figure 12. It is evident that the public health use of DDT will centre on the equatorial zone with lesser applications extending up to 45° latitude in either hemisphere. Thus, one notes a displacement of potential pollution by DDT residues from the mid-latitudes of the northern hemisphere, where the bulk of the DDT has been produced, towards the south where the DDT will be used.

The predictions of Rafatjah and Stiles are based upon present use data. The actual uses could be higher or lower. Of the 1,827 million people living in malarial areas, 450 million are currently being protected by antimalarial measures.

There are still 440 million people in malaria areas who have not yet been protected against the disease through the use of DDT. This population is expected to be included in future programmes, which would result in increased use of DDT. No economical substitute for DDT has yet been developed. Possible reversions in advanced malaria eradication programmes could require increased use of DDT.

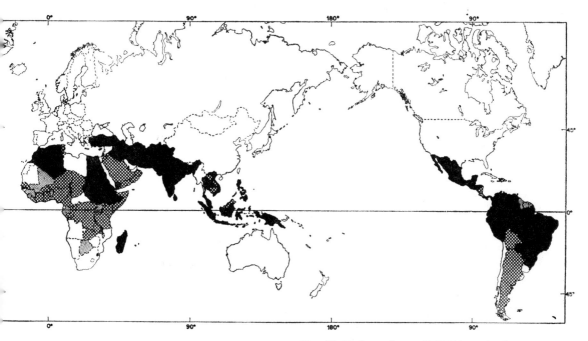

FIG. 12. Estimated use of DDT in antimalaria programmes for 1972–81 (Rafatjah and Stiles, 1972). Black, greater than 10,000 tons; hatched, between 1,000 and 10,000 tons; stippled, less than 1,000 tons.

TABLE 6. Estimated DDT requirements for antimalaria programmes (in tons)

WHO region	Formulation*	1971	1972	1973	1974	1975	1976	1977	1978	1979	1980	1981	Total 1971–81
African	Technical	62	80.4	90.4	100.4	113.4	113.9	112.1	112.1	112.1	112.1	112.1	1,059
	75 per cent wdp	632.4	655.4	697.4	791	815.8	828.5	834.1	884.1	884.1	884.1	884.1	8,208.6
	25 per cent wdp	100	120	133	145	159	175	166.4	166.4	166.4	166.4	166.4	1,564
American	Technical	557.1	557.5	649.2	601.4	559.7	517	517	504.5	431.2	429.7	353.7	5,120.9
	75 per cent wdp	11,286	11,328.3	11,120.6	10,427.4	9,697	8,624.9	8,424.9	7,724.9	6,758.5	6,563.5	5,325	85,995
South-East	75 per cent wdp	16,435.3	21,424.5	20,607.9	21,578	22,387.8	25,039.9	30,983	26,348.3	17,765.7	18,312.2	12,709.9	217,157.2
Asian	50 per cent wdp	8,000	6,000	6,000	6,000	6,000	5,400	6,000	6,000	6,000	6,000	6,000	59,400
European	75 per cent wdp	700	980	980	880	880	880	200	200	200	200	200	5,600
	50 per cent wdp	500	500	500	500	500	300	300	200	300	300	300	3,800
Eastern	Technical	31	31	31	31	31	31	31	31	31	31	31	310
Mediterranean	75 per cent wdp	14,090	9,084	8,614	8,454	8,454	8,106	7,936	7,426	7,211	7,171	7,171	79,627
Western	75 per cent wdp	1,235	1,962	2,360	2,438	2,515	3,028	3,375	3,215	2,785	2,610	2,390	26,678
Pacific	25 per cent EC	365	367	524	724	715	715	715	715	515	515	515	6,020
Total (tons)	Technical	650.1	668.9	770.6	732.8	704.1	661.9	660.1	647.6	574.3	572.8	496.8	6,489.9
Total (tons)	75 per cent wdp	44,378.7	45,434.2	44,379.9	44,568.4	44,749.6	46,507.3	51,803	45,798.3	35,604.3	35,740.8	28,680	423,265.8
Total (tons)	50 per cent wdp	8,500	6,500	6,500	6,500	6,500	5,700	6,300	6,300	6,300	6,300	6,300	63,200
Total (tons)	25 per cent { wdp / ec }	465	487	657	869	874	890	881.4	881.4	681.4	681.4	681.4	7,584

* wdp, water-dispersible powder; ec, emulsion concentrate.
Reproduced from H. A. Rafatjah and A. R. Stiles, *Summary Review of the Use and Offtake of DDT in Antimalaria Control Programs*, Geneva, WHO, 1972. (VBC/72.5.)

Thus, it appears that the present level of use, or slightly smaller amounts as predicted by Rafatjah and Stiles, will continue even beyond the 1980s.

PCBS

These are the chlorinated isomers of biphenyl. There are more than 200 theoretically possible compounds of the type where 'x'

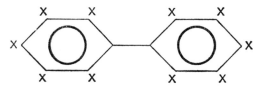

indicates possible chlorine positions. The commercial products usually contain between 40 and 60 per cent chlorine by weight. (In the United States the Monsanto PCBs known as Aroclors had their chlorine percentage indicated in the last two numbers of their product number. Thus, Aroclor 1254 is a polychlorinated biphenyl containing 54 per cent chlorine by weight.) They are prepared by the chlorination of biphenyl which results in the production of mixtures of different polychlorinated biphenyls, difficult to separate. The PCBs used in industry probably consist of at least eighty different substances. Since individual chlorobiphenyls may react differently in the environment and may affect organisms differently, a knowledge of the exact constituents in the starting product and in different reservoirs is needed. Techniques for their separation and analysis have been devised by Jensen and Sundstrom (1974).

USES

The industrial uses depend upon the unique chemical characteristics of the PCBs: high stability, non-flammability, low water solubility, low volatility and high dielectric constant. Their utilizations fall into two categories. The first includes those applications in which the PCBs are used in closed systems and are recoverable. If they are partially altered chemically, there is the possibility of regenerating the useful materials. The second includes those dissipative uses in which there is a very high or in some cases total loss to the environment during or following its utilization.

In the first category, PCB use as dielectrics for transformers and for large capacitors has been extensive. Theoretically, effective engineering design of the units can minimize the escape of PCBs.

In the second category the uses of PCBs in hydraulic fluids, especially in systems involving high temperatures, and as heat-transfer fluids are also widespread, but usually entail small amounts of the chemicals. There are great difficulties in recovering small quantities of these chemicals from a large number of users. Therefore such uses are primarily dissipative.

In addition, PCBs have been used as plasticizers, lubricating and cutting oils, components of paints and printing inks, sealants and resin extenders in adhesives. Following these uses the PCBs are essentially unrecoverable.

In the United States, before restrictions of sales to only non-dispersive applications in 1970, 60 per cent of the industrial uses were for closed-system electric and heat-transfer equipment, 25 per cent for plasticizers, 10 per cent for hydraulic fluids and lubricants and less than 5 per cent for other dispersive uses.

PCB PRODUCTION

Like DDT, world-wide production data for PCBs are not available as a function of time. The data for the United States, provided by the Monsanto Corporation, the sole producer in the United States, goes back to 1960 (Table 7). OECD (1973) gathered production data for its member countries

TABLE 7. Production of PCBs in the United States and Japan (in kilotons per year)

Year	PCB production	
	United States*	Japan**
1971	18.4	6.8
1970	38.6	11.1
1969	34.6	7.7
1968	37.6	5.1
1967	34.2	4.5
1966	30.0	4.4
1965	27.4	3
1964	23.1	2.7
1963	20.3	1.8
1962	19.0	2.2
1960	18.9	2.2
1959		1.6
1958		0.88
1957		0.87
1956		0.5
1955		0.45
1954		0.2

* Data from Monsanto Corporation.
** Data from K. Fujiwara, *Environmental and Food Contamination with PCBs in Japan*, Department of Hygienic Chemistry, Kyoto City Institute of Public Hygiene, 1974.

TABLE 8. Production of PCBs in certain countries, 1971 (in tons)

Country	Quantity
United States	18,000*
Federal Republic of Germany	8,000
France	7,600
United Kingdom	5,000
Japan	6,800**
Italy	1,500
Spain	1,500•
OECD countries	48,000

* 38,000 tons in 1970.
** 11,000 tons in 1970.
• Estimate.

Reproduced from OECD, *Polychlorinated Biphenyls, their Use and Control*, Paris, OECD, 1973, 42 p.

for the year 1971 (Table 8). More detailed Japanese data are given in Table 7 Japanese production was about 10 per cent of United States production in the 1960s. Japan has a surface area about one-thirtieth that of the United States. Thus they are liable to greater pollution problems if relative leakages to the environment are similar in both countries.

There was a decided drop in the United States production of PCBs between 1970 and 1971. This was due to the recognition by Monsanto that this collective of chemicals may threaten environmental resources. In August of 1970, the Corporation discontinued all sales for general plasticizer application where environmental leakage was high. In addition, they phased out over a six-month period sales to users of PCBs as components of industrial and hydraulic fluids where control of spillage and collection of spent material was not possible. Sales of the chemicals as dielectric fluids for transformers and capacitors, and as fire-resistant heat-transfer fluids, were continued. The Monsanto Corporation cooperates with customers in the collection of the spent fluids for regeneration and/or incineration.

Manufacturers in other countries have also limited sales or ceased production (OECD, 1973).[1] Japan suspended manufacture in June 1972. The United Kingdom and Federal Republic of Germany have restricted sales to uses regarded as nonpolluting (such as in dielectric, heat-transfer or hydraulic fluids). In France, one manufacturer has discontinued sales of PCBs for use as cutting oils, for heat-transfer fluids in pharmaceutical and food industries, in the production of carbonless copying paper and in marine paints.

1. OECD actions recommending to member countries to limit PCB usage to non-polluting uses were endorsed by WHO in January 1974 and this information was sent to its member countries.

In three non-producer countries (Norway, Sweden and Switzerland) legislation restricts the use of PCBs by limiting it to essentially closed systems, whereas the governments of the Netherlands and Australia have used persuasion to limit sales to non-dispersive uses.

PCBs and DDT residues in the ocean system

FLUXES OF DDT AND PCBS TO THE OCEANS

The initial flux calculation of DDT to the oceans (NAS, 1971) was made by utilizing the measured values of DDT in rains and an estimated total annual precipitation of water over the oceans estimated to be 3.0×10^{20} cm^3 per year. At the time the report was written only two sets of rainfall data were available to the authors: rainfall in the United Kingdom, measured at seven stations between August 1966 and July 1967, with a mean DDT concentration of 80 parts per trillion (p.p.t.) (Tarrant and Tatton, 1968); and south Florida precipitation taken at 1,000 parts of DDT per trillion (Yates *et al.*, 1970). NAS (1971) used the former, more conservative value in their flux calculation and estimated a potential injection of 2.4×10^{10} grammes of DDT per year into the marine system. At the time this was about one-quarter of the estimated annual world-wide production of DDT. This computation indicates that the troposphere must be a major route for the transfer of DDT residues into the oceans.

This calculation has been strongly criticized as a global extrapolation from a single set of data. However, an examination of more recently reported rainfall data suggests that the 80 p.p.t. value[1] may have provided a reasonable estimate of DDT movement in the atmosphere during the period of high usage in the northern

hemisphere. The data for the compound p, p' DDT was: in Hawaii (1970–71) a concentration of 1–13 p.p.t. (i.e. an average of 5) (Bevenue *et al.*, 1972); in Ohio (1965) a concentration of 70–340 p.p.t. (i.e. an average of 187) (Bevenue *et al.*, 1972); in central England (1965) a concentration of 2–4 p.p.t. (i.e. an average of 3) (Bevenue *et al.*, 1972); and, finally, in the Federal Republic of Germany (1970–72) a concentration of 1–100 p.p.t. (Weil *et al.*, 1973).

The higher values in Hawaii and in the Federal Republic of Germany in the 1970s after curtailment in use do emphasize the importance of the continuing flux of DDT to the environment. The Hawaii figures are especially relevant inasmuch as they most probably reflect transoceanic as well as transcontinental movements of DDT. Further, these numbers are probably somewhat low since the often more abundant DDE is not included.

The rainfall calculations have another uncertainty associated with them—the effectiveness of rain in scavenging the DDT residues and the PCBs from the atmosphere, be they associated with particulates or in the vapour state. Stanley *et al.* (1971) found in an agricultural area that the pesticide levels could be better correlated with reported spraying than with rain. On the other hand, Ware and Addison (1973) found a significant linear correlation between PCB levels in the Gulf of St Lawrence plankton and the cumulative amount of rainfall ten to twenty days before sampling. Still, the ability of precipitation to remove these halogenated hydrocarbons from the atmosphere remains to be established.

1. It is difficult to assess the representativeness of the few published rainfall data. Clearly, some may be related to local usage; others to the history of the air-masses through which the rain fell. None the less, with the existing data it is difficult to improve upon this approximative calculation.

The sediments of the Santa Barbara basin provided a historical record of PCB and DDE fluxes to the coastal ocean from the Los Angeles area (Hom *et al.*, 1974). The period of 1890 to 1967 was examined. DDE first appears in measurable concentrations in 1952, whereas PCB deposition began earlier, about 1945, following its rapid increase in use during the Second World War as a dielectric fluid, as a paint additive and in sundry other ways. Up to the end of 1967 the rates of deposition for these two groups of compounds increased.

In 1967, the estimated annual fluxes to the Santa Barbara basin were 1.9×10^{-4} g/m^2 and 1.2×10^{-4} g/m^2 for the DDE and PCBs, respectively. These fluxes are about twice those calculated in the NAS (1971) report (2.4×10^{10} grammes of DDT per year is equivalent to an annual deposition rate of 8×10^{-5} g/m^2). The agreement is satisfying since the flux based on rainfall in the United Kingdom represents atmospheric washout whereas the California figures encompass those substances introduced to the sediments by all transport paths.

These fluxes for 1967, if applicable to other regions of the coastal area near Los Angeles, would give a total annual deposition of 10 tons of DDT residues over an area 200 kilometres by 200 kilometres. The source of this DDT is attributed to municipal waste waters which received their DDT burden from a factory. In 1971 19 tons of DDT were transported to the waters off Los Angeles. DDT in drainage waters from agricultural areas are generally much lower. Only 2 tons in 1967 are estimated to have entered San Francisco Bay which drains one of the principal agricultural areas of California (Hom *et al.*, 1974).

The PCB flux for this 200 kilometres by 200 kilometres area, based on the Santa Barbara results, would be 4 tons per year. The annual input of PCBs into the region from waste-water outfalls and from river runoff in 1970–71 has been estimated

at 10 and 0.25 tons per year, respectively.

An important observation from these studies is that a substantial portion of the chlorinated hydrocarbons entering the sea is rapidly deposited in coastal sediments. In addition, because of the anoxic nature of these deposits and the lack of disturbances in them by burrowing organisms, a record of the changes of flux with time is recorded within them.

Suspended particulates in coastal waters may be especially effective in scavenging any dissolved DDT residues and transporting them to the sediments. Adsorption may occur on both the organic and mineral components (Pierce *et al.*, 1974). Humic-type particulates have a greater adsorption capacity for DDT than do clay minerals or other components of marine sediments, according to Pierce and his co-workers. Continental sources of the pesticides, including soil-erosion debris and sewage effluents may contain large amounts of humic substances. The DDT residues may enter the marine environment already sorbed to such particles. A mechanism is thus provided for their uptake by filter- and detritus-feeding organisms and entry to the food web.

PCB AND DDT
ENVIRONMENTAL LEVELS

Atmospheric chlorinated hydrocarbons (PCBs and DDT) in marine air collected in the Bermuda-Sargasso Sea area of the North Atlantic appear to be in the gas phase according to the results of experiments conducted in February to June 1973 (Bidleman and Olney, 1974). Less than 5 per cent of the combined total of these compounds could be captured by a glass-fibre filter capable of retaining particles with radii greater than 0.015 μm.

The authors point out that there is the possibility that the organochlorine compounds have volatilized off the particulates captured by the glass-fibre filter. Previous workers had collected particulates in similar ways. These experiments did lead to finding DDT values two orders of magnitude greater than those of previous workers who merely analysed the solid phases. Since DDE evaporates more readily than DDT, perhaps an analysis of the DDE/DDT ratios on the filter and in the gas phase would have revealed how extensive such a volatilization process might have been.

Bidleman and Olney (1974) noted that the PCBs in the air matched in composition Arochlor 1242 or Arochlor 1248 (i.e. PCBs with 42 or 48 per cent chlorine) while those extracted from the Sargasso Sea resembled Arochlor 1254 or 1260 in chlorine content. Therefore, either the more highly chlorinated biphenyls are selectively transported to the oceans or the less-chlorinated biphenyls are more readily degraded in the water and more rapidly evaporated from water.

Concentrations of both groups of compounds were higher in the surface monolayer (15 μm) than in the subsurface waters. Perhaps this results from an adsorption of these chlorinated molecules by natural and synthetic organic chemical slicks.

The average residence times of these compounds were estimated for the airs above the Sargasso Sea waters by Bidleman and Olney (1974) with the following assumptions:

The compounds are uniformly distributed in the mixed layer to a depth of 100 metres with average concentrations for PCBs of 1.1×10^{-6} g/m^3 and for DDT residues of 6×10^{-8} g/m^3.

The residence time in the mixed layer is 4 years.

The atmospheric concentrations are uniform up to the tropopause (6.3 km of air equivalent height) with an average concentration of PCBs of 5×10^{-10} g/m^3.

The major route of PCBs and DDT into the ocean is through the atmosphere.

There is no sea to air transport.[1]

The flux of PCBs leaving each square metre of mixed layer to the deeper waters is equal to that entering the mixed layer from the atmosphere and is given by F=amount in the mixed layer/residence time=100 m$\times$ 1 m$^2 \times 1.1 \times 10^{-6}/4 = 2.8 \times 10^{-5}$ grammes per year.

Since this is the flux out of the atmosphere, the atmospheric residence time is given by: $T = 6.3 \times 10^3 \times 5 \times 10^{-10}/2.8 \times 10^{15}$ =0.11 year or about 40 days.

A similar calculation gives a residence time for DDT of 51 days.

Higher concentrations of PCBs were reported by Harvey *et al.* (1974a, 1974b) in eastern and western North Atlantic waters for the years 1971 and 1972. Surface waters averaged about 30 ng/l. By April, 1973, the PCB concentrations in the Sargasso Sea dropped to 1 ng/l. The decline continued, according to Harvey *et al.* (1974a), until February 1974 when 0.8 ng/l were measured. Such losses would require very rapid removal rates. Harvey *et al.* (1974a) suggest that the behaviour of plutonium isotopes in the marine environment, following entry from the atmosphere, provides an explanation for the phenomenon. The rapid removal of plutonium isotopes from surface waters was explained by their association with particles falling at rates of 70–392 m/year. A 1973 profile of PCBs at a station in the Atlantic (32°25′ N.; 70°20′ W.) indicates a wide dispersal in the water column (see Table 9).

Other processes could lead to a short residence time of PCBs in surface waters: evaporative co-distillation or biogeochemical degradation. The importance of either has not as yet been established. Harvey *et al.* (1974a) relate to the rapid change

1. The transfer of these halogenated hydrocarbons from water to air may be much greater than is appreciated at present on the basis of kinetic equations developed to predict the rate of evaporation of organic compounds from water (Mackay and Wolkoff, 1973).

Table 9.

Depth (m)	PCB concentration (ng/l)	Depth (m)	PCB concentration (ng/l)
10	0.6	600	0.5
100	0.8	900	1.9
300	0.9	5,100	0.4

in surface-water concentrations to the reduction in the industrial discharges of
PCBs.

Two other characteristics of the PCB distributions found in the above studies are significant. There was a wide range in surface-
water concentrations (<1 to 150 ng/kg)
taken during the same time period. For
example, two surface-water samples 80 kilometres apart, had PCB concentrations of 2
and 9 ng/kg. Such a variability, if real,
may be accounted for by a preferential
uptake of the PCBs by sea slicks, by localized rainfalls sweeping the PCBs from the
atmosphere or by discharges from ships.

One of the most interesting observations
was that the ratio of DDT/PCB was found
to be less than 0.05 in these surface waters.
Similar ratios were observed by Bidleman
and Olney in both surface microlayer and

subsurface waters. Yet, on the basis of production data (Tables 3 and 5), the ratio is
expected to be substantially greater than 1.
If losses and transport paths to the oceanic
environment are similar, then it appears
that the PCBs have a far greater stability
than DDT and its metabolites.

LEVELS IN MARINE ORGANISMS

DDT residues and PCBs are associated with
the lipid fraction of marine organisms. The
particular level in any single organism or
group of organisms depends upon their exposure histories—i.e. the levels of these
pollutants in the waters in which they live
and in the food they consume.

DDT residues and PCBs have been
measured in a variety of plankton samples
(IDOE, 1972; Harvey *et al.*, 1974*b*; Williams
and Holden, 1973; Ware and Addison,
1973) (see Table 10).

There was a marked gradient in the
PCB and DDT residues concentrations in
plankton samples taken from the Scottish
coast going from the Firth of Clyde to an
open-ocean station 400 miles west. DDT
residues decreased by a factor of ten and
PCBs by a factor of twelve (Williams and
Holden, 1973). The source of the PCBs
closer to the coast was primarily sewage

Table 10.

Area	No. of samples	DDT residues	PCBs
		p.p.b. wet weight	
Gulf of St Lawrence	9		2,000–93,000
Scottish coast	26	<3–107	10–2,200
Sargasso Sea	4	0.7	7–450
South Atlantic	4	0.2–2.6	19–638
North-East Atlantic	22	2–26	10–110
Clyde, Scotland	15	6–130	40–230
California, United States		0.2–206	0.7–30
California, United States			100–1,300
Iceland (phytoplankton)	1		1,500
North and South Atlantic	53		0.2

sludges from the city of Glasgow (Halcrow et al., 1974) dumped in the Firth. PCB levels in superficial sediments indicate that these chlorinated hydrocarbons have been dumped for about twenty years with a marked increase about ten years ago. The benthic fauna in the area are highly contaminated with PCBs.

The plankton samples from the Gulf of St Lawrence had PCB concentrations that were inversely related to particle size (Ware and Addison, 1973). The plankton, collected with a No. 20 net, ranged in size from 73 to 1,050 μm. This observation is in accord with the affinity of these halogenated hydrocarbons for particulate matter. Atmospheric input was postulated to be the likely source of supply.

The last set of the above analyses was carried out by Harvey et al. (1974b) on samples collected between December 1970 and August 1972. The four samples containing more than 1 p.p.b. by weight were very rich in phytoplankton. In all samples the PCB/DDT residue ratio was always greater than thirty as in surface waters. Both the North Atlantic and South Atlantic samples had the same average concentrations, a result still unexplainable. The vertically migrating mesopelagic organisms in the Atlantic Ocean, which live in the 300–1,000 m water depths, contained 1–170 μg/kg of PCBs and 0.5–29 μg/kg of DDT residues on a wet weight basis. There was no covariance between the halogenated hydrocarbons concentrations in the organism and depth of habitat, lipid content, migratory habits, age or trophic level. Many of the organisms were fish. Harvey et al. (1974b) suggest that for the chlorinated hydrocarbons the lipid fractions of gilled organisms are in some sort of chemical equilibrium with the sea water and that the pollutants enter the blood lipids through permeable body surfaces, such as the gills. The chlorinated hydrocarbons would then be partitioned between blood lipids, and

presumably by circulation, with other body lipids and the sea water directly through permeable body membranes.[1]

A long-term surveillance of DDT and its degradation products in estuarine molluscs was undertaken by the Gulf Breeze Laboratory (now a part of the United States Environmental Protection Agency). The period 1965–72 was covered and oysters, mussels and clams were used as assay organisms (Butler, 1973). Laboratory experiments with the eastern oyster *Crassostrea virginica* indicated that its uptake and flushing rates made it, as well as other molluscs, useful indicator animals. The oysters could accumulate DDT from ambient waters containing levels as low as 10 p.p.t. At a DDT water concentration of 1 p.p.b. they could take up as much as 25 p.p.m. of DDT in their tissues following a 96-hour exposure. In addition the oysters can flush DDT from their systems rapidly after exposure to uncontaminated waters. For example, DDT residues of 25 p.p.b. in oysters were decreased by 50 to 90 per cent following a week of flushing.

The results from fifteen coastal states of the United States (Butler, 1973) showed DDT to be measurable in 63 per cent of the samples. Of the 7,000 analyses, the mean residue composition was 24 per cent DDT, 39 per cent DDD and 37 per cent DDE.

The conclusion was reached that DDT pollution in most estuaries peaked in 1968 and has been declining markedly since 1970. The number of samples that contained less than 11 p.p.b. of DDT was 76 per cent of the total in 1971 compared with 8 per cent in 1966 and 1967. This is illustrated in Figure 13 for ten stations off North Carolina, where the percentage of oysters containing more than 10 p.p.b. of DDT

---

1. Such a mechanism will not explain the high concentration factors for organohalides in seals (Williams and Holden, 1973).

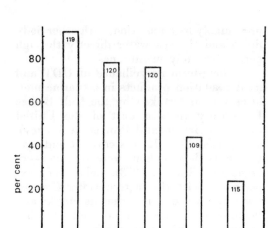

Fig. 13. Percentage of eastern oyster samples, collected at ten stations in North Carolina, containing more than 10 p.p.b. of DDT. Numbers in bars indicate total number of samples (Butler, 1973).

steadily decreased between 1967 and 1971.

A surveillance over the years 1969–71 of the DDT levels in mussels (Canada, Denmark, Finland, Netherlands, Norway, Spain, Portugal, Sweden, United Kingdom and the United States) disclosed that the highest DDT residues were in samples from the United Kingdom and the Mediterranean coast of Spain (Holden, 1973). The concentration ranged from less than 0.01 to 30 p.p.m. wet weight. Herring, analysed from the same areas (except Norway), sometimes had higher amounts of DDT residues than those in the molluscs. Highest DDT values were from the Baltic.

Food-chain magnification of PCBs and DDT was not observed by Harvey *et al.* (1974*b*), especially among gilled organisms. Analyses were made of Atlantic biota collected between December 1970 and August 1972 from the open Atlantic Ocean between 66° N. and 35° S. All organisms collected had higher concentrations of PCBs than of DDT residues. Plankton had the highest concentrations of PCBs while flying fish which feed mainly on plankton carried

a body burden of the PCBs an order of magnitude less.

Harvey *et al.* (1974*b*) found no discernible concentration gradients of PCBs or DDT residues in organisms from near shore to the open sea, probably as a result of a lack of sampling in the coastal margins. Williams and Holden (1973) observed PCB concentration gradients for plankton going out from the coast of Scotland. Giam *et al.* (1972) determined DDT, DDE, and PCBs in biota from the Gulf of Mexico and the Caribbean Sea in 1971 and noted that samples from coastal areas generally had higher levels than samples from open waters. Probably, for any given coastal area and its adjoining open ocean waters, the concentration gradients in the biota will reflect a variety of factors—wind patterns, current systems, biological productivity, industrial and agricultural activities on the adjacent land, and temperature régimes.

In biota from the Antarctic, chlorinated hydrocarbons were detected in all five samples analysed (decapod larvae, bulk plankton, euphausiids, sponges and fish), but the levels were low (p.p.b. wet weight or less). These concentrations were two orders of magnitude lower than those in open-ocean biota from the Gulf of Mexico and the Caribbean Sea (Giam *et al.*, 1973).

From this rather extensive set of analysed samples, another facet of the environmental chemistries of these halogenated hydrocarbons emerged. Although similar in molecular structure, the PCBs and DDT residues have quite different paths within the ocean system. The PCB/DDT residue ratio is about thirty in the marine atmosphere, surface waters and plankton, and decreases to three in higher predators in the food web and in mid-water organisms.

ENVIRONMENTAL IMPACTS

Laboratory experiments suggest that the halogenated hydrocarbons may influence the makeup of phytoplankton communities by preferential impact upon one species. The concentrations used in the experimental work are several orders of magnitude greater than the environmental levels usually measured. However, the possibility of unusually high concentrations in localized areas is possible, such as those near sites of production or use, or resulting from an accidental discharge.

Mosser *et al.* (1972) found that PCBs inhibited the growth of the diatom *Thalassiosira pseudonana* in pure culture at concentrations of 25 p.p.b. but not at 10 p.p.b. Growth was diminished by DDT at 100 p.p.b., inhibited at 50 p.p.b. and not affected at 25 p.p.b. in similar experiments. The green alga *Dunaliella tertiolecta* in pure culture was not affected by these chemicals at the concentrations used in the above experiments.

In mixed cultures, exposure to either PCBs or DDT enhanced the population of *D. tertiolecta* relative to that of *T. pseudonana*. In control cultures the diatom grew faster and became dominant. Since many species of zooplankton preferentially graze upon specific species of phytoplankton, the health and distribution of organisms all the way up the food chain could similarly be affected.

Juvenile shrimp, when exposed to 5 p.p.b. of the PCB Arochlor 1254 in flowing sea water, experienced a 75 per cent mortality within twenty days (Duke *et al.*, 1970). These experiments were carried out in response to an industrial leakage of PCBs into Escambia Bay, Florida, where the waters contained less than 1 p.p.b. of PCBs and the shrimp 2.5 p.p.m. (on a wet weight basis). Both levels are sublethal ones. Juvenile blue crabs were less sensitive to a 5 p.p.b. sea water exposure with only one of twenty crabs dying during the twenty-day exposure. An average whole-body residue in five of the crabs was 23 p.p.m. (wet weight) with a range of 18

to 27 p.p.m. The persistence of the PCBs in animal tissues is indicated by a subsequent experiment in which six crabs were held in uncontaminated waters for four weeks following exposure. The average body burden only dropped from 23 to 11 p.p.m.

The ecological significance of such laboratory studies is not known inasmuch as the techniques to measure population alterations (in the absence of mass mortalities) have not yet evolved to the point of application. None the less, results of this type indicate that population changes might take place and that the development of techniques to detect such change is desirable.

Disturbances of the routines of marine organisms in their normal habitat would appear to be a more convincing way to demonstrate the biological effect of a pollutant. However, the difficulties of establishing cause and effect relationships are well recognized. Perhaps, the problems associated with the following investigations can illustrate this dilemma.

Premature pupping in sea lions breeding in California's Channel Islands had been noted in 1968 and in following years. Delong *et al.* (1973) sought to determine whether environmental pollutants might be responsible and directed their studies to the halogenated hydrocarbons as the possible toxins. Organochlorine pesticides and PCBs were found in all tissues from the mothers and from the pups, and the most common chemical was found to be DDE (80–93 per cent of the total DDT). All tissues from females that gave birth to premature pups in 1970 contained higher concentrations of these chemicals than did the tissues from females that carried their pups to full term. The DDT residues in the blubber and liver of prematurely parturient females were 8 and 3.8 times greater, respectively, than those of normal parturient cows. There were twice as much DDT residues in the brains of premature pups as in those of full-term pups. Similar relationships were

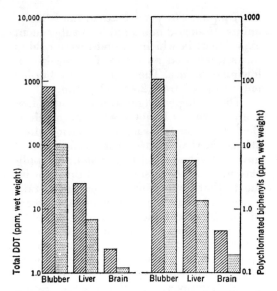

FIG. 14. Mean concentrations of total DDT and PCBs in tissues of California sea lions delivering premature (cross hatched) and full term (stippled) pups. From Delong *et al.*, 1973. (Copyright 1973 by the American Association for the Advancement of Science.)

noted in the PCBs. For both collectives of compounds, there was no overlap of concentrations between members of the premature and full-term groups. The data are summarized in Figure 14.

Delong *et al.* (1973) provide an explanation for the two populations of mothers. There may be two groups of sea lions utilizing different feeding areas. Some females are believed to move southward and winter in waters off southern Baja California while others remain in the general area of the Channel Islands. Organochlorine residues in marine organisms in the latter area are known to be higher than in the former zone. The Channel Islands population would be exposed to a higher intake of the compounds in their food.

It is tempting to relate the premature pupping to organochlorine compounds. At this time, however, the case is not strong. First of all, could this be due to another pollutant acting alone or in combination with the halogenated substances? The authors analysed mercury in the premature parturient females and in their pups, but not in the group of mothers who gave birth to full-term pups or in these pups.

In addition, the cause-effect relationship would be strengthened with a knowledge of the biochemical or physiological relationship between premature births and the pollutant.

A somewhat more substantial cause-effect relationship seems to exist between DDE in birds and the thinning of the shells of their eggs (Blus *et al.*, 1972). Determinations of DDE concentrations (DDE is the predominant DDT residue in these studies) were made upon eighty eggs of brown pelicans (*Pelecanus occidentalis*) collected in 1971 and 1972, seventy in 1969 from twelve colonies (one in California, two in South Carolina and nine in Florida) and ten more eggs collected in 1970 from one of the South Carolina colonies.

There is a logarithmic relationship between the amount of egg-shell thinning and the concentration of the DDE in the eggs (Fig. 15). The percentage of thinning was based on the pre-1947 mean-thickness of the eggs as measured by earlier investigators. A relationship of the type: $Y = B - bX$ describes the data, where Y is the percentage of the average pre-1947 egg-shell thickness, B is the expected percentage thickness at 1 p.p.m. of DDE, X is the logarithm of the content of DDE in the egg on a fresh wet weight basis in p.p.m., and b is the regression coefficient.

Both DDT and its residues and the PCBs have been implicated in the reproductive failures, primarily due to egg-shell thinning of marine birds and these results may assist in the understanding of the population decreases. As a consequence of the thinning, the friable eggs are subject to damage by parental pecking and by other

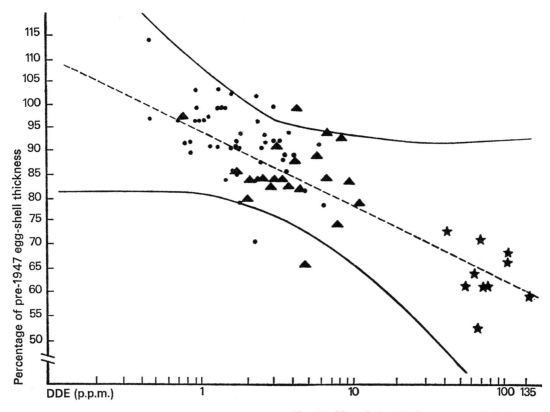

FIG. 15. The relationship between egg-shell thinning and DDE in eighty brown pelican eggs. Solid lines represent 95 per cent confidence limits (Blus *et al.*, 1972).

●, Eggs from Florida pelicans; ▲, eggs from South Carolina pelicans; ★, eggs from California pelicans.

means. A 15 per cent thinning of the pelican eggs is associated with DDE residues between 4 and 5 p.p.m. In the South Carolina populations which were on the decline, eight of the eleven eggs in 1969 and two of the ten collected in 1970 contained DDE in excess of 4 p.p.m.

The brown pelican population of Anacapa Island off the California coast, the same colony as in the above study, experienced severe population decreases between 1969 and 1972 which have been attributed to egg-shell thinning (Risebrough, 1972). A 35 per cent decrease in egg-shell thickness was associated with 71 p.p.m. of DDE in the eggs. The source of the DDT residues was from a factory in Los Angeles which discharged 200–500 kilograms of DDT daily into the Los Angeles sewage system from where it emptied into the sea. The DDT residues were accumulated in marine organisms which were subsequently eaten by the pelicans and presumably effected the egg-shell thinning.

EFFECTS ON HUMAN HEALTH

The ingestion of foods containing present or past environmental levels of DDT has had no discernible effect upon human health. Human deaths and morbidity, on the other hand, have been associated with the consumption of PCBs (Kuratsune *et al.*, 1972). In 1968 a Japanese rice oil was contaminated during its manufacture by the leakage of PCBs from heat exchangers. In some cases, 0.5 to 2.0 grammes of PCBs per person were consumed as a consequence of the oil's use in the preparation of foods. More than 5,000 people suffered from a chloracne-like skin eruption. In addition, some of those who consumed the contaminated oil came down with palsy, fatigue, and vomiting spells. The deaths of two adults and of two babies were attributed to the intake of these chemicals. Several babies were born with abnormal pigmentation of the skin. There have been no other reports of PCB impacts upon human health and there is no evidence that present PCB body burdens pose any threats to man.

DDT TRANSFORMATION

The major metabolite of DDT excreted from marine organisms is DDD (Patil *et al.*, 1972). Transformation of DDT took place in biologically active samples such as surface films, plankton, algae and sediments. However, the insecticide is not readily metabolized in plain sea water, even from such polluted estuaries as Kaneohe Bay and Pearl Harbor, both in Hawaii. One hundred microbial isolates from Hawaii and Houston were investigated to test their abilities in degrading DDT, thirty-five of which appeared to be active.

In laboratory and field studies of DDT breakdown in the Severn estuary (one of the major estuarine regions of the United Kingdom, draining one-sixth of the total land area of England and Wales), the prin-

cipal metabolite was DDD (Albone *et al.*, 1972).

When radio-active DDT was suspended in sea-water samples maintained in the laboratory, small quantities of DDD were produced. A greater conversion of DDT to DDD occurred when the samples were placed in a hydrogen atmosphere.

The greatest DDT breakdown occurred where the DDT was incubated in anaerobic sewage sludge, again in hydrogen. Bacteria isolated from the estuarine muds were often able to degrade DDT *in vitro*.

THE VAPORIZATION PROCESS

Following application as a pesticide in agricultural or public-health activities, DDT and its degradation products enter the environment primarily through vaporization. Field, laboratory and kinetic investigations have revealed the details of this process.

Of consequence is the observation that the vapour pressure of DDE is eight times greater than that of its parent; the saturation vapour pressure for DDE is 109 ng/l as compared to 13.6 ng/l for DDT at 30° C (Spencer and Cliath, 1972). In soils or on a surface DDT is converted to DDE or other degradation products by microbial or photochemical decompositions. The higher volatility of DDE results in a marked fractionation of it from its parent between the soil or surface and the atmosphere. This is verified in the experimental work of Cliath and Spencer (1972) who measured the DDT residues in a soil, which had been treated fourteen months previously, and the atmosphere above the soil. DDT was sprayed ten to twelve times per year in amounts of 3 pounds per acre upon sweet corn for earthworm control in Cuachilla Valley, California. The results are given in Table 11.

The concentrations of the DDE, relative to the DDT, were higher in the atmosphere than in the soil. The field data give a picture of the fate of DDT which is probably

TABLE 11.

Chemical	In soil (µg/g)	In atmosphere (ng/m³)
p, p' DDT	14.1	91
p, p' DDE	6.2	327
o, p DDT	2.7	0.61
o, p DDE	0.12	0.24

quite different from that derived from short-term volatility experiments in the laboratory.

The rates of volatilization of DDT and its degradation product have been approached both theoretically and experimentally in the laboratory. The rates of vaporization are dependent upon their vapour pressure and upon the rate of diffusion of evaporated molecules through the layer of still air above the deposit (Lloyd-Jones, 1971).

Using a vapour pressure of DDT equal to 1.5×10^{-7} mm of mercury at 20° C, a diffusion coefficent of 0.05 cm²/s and a still-air thickness of 2 mm, an evaporation rate of 3×10^{-3} µg/cm²/h is computed for 20° C. A second method, based upon the kinetics of the evaporation process and using the relationship between the loss rate and the product of the vapour pressure and the square root of the molecular weight, gave a value of 2×10^{-3} µg/cm²/h (Lloyd-Jones, 1971).

Lloyd-Jones also sought experimentally the evaporization rate using DDT labelled with carbon-14. The DDT was placed upon plastic rings or screens for periods of forty-three to fifty-three days and the DDT loss was ascertained. Measured rates varied between 2×10^{-3} and 3×10^{-4} µg/cm²/h, in agreement with the calculated values.

If the amount of DDT produced annually in the world today (100,000 tons) were to evaporate in one year from a sur-face area equivalent to that of the earth's lands, 2×10^{18} cm², an evaporation rate of 5×10^{-8} g/cm² would be required for a single evaporation process. The lowest rate measured by Lloyd-Jones is equivalent to 2.6×10^{-6} g/cm² per year. Thus, these measured or calculated rates can clearly account for the vaporization of the annual production of DDT.

The vaporization of DDT from soils or from surfaces can be enhanced by the presence of water. Two mechanisms are operative and have been described by Hartley (1969). The first he calls 'wick evaporation'. When vaporization takes place from the end of a wick, the lost solution is replaced by that which moves up through the wick interstices. If the solute is less volatile than the liquid, it will concentrate at the site of evaporation. Hartley suggests the following three consequences of such a train of events. First of all, the vaporization of water can be reduced as the solute attains higher and higher concentrations. Or the solute can reach a steady-state concentration in which its rate of vaporization is compensated by the arriving amount. Finally, the solute can precipitate, if its solubility is exceeded. In the solid state it is more readily volatilized. Any or all three may occur in a given system.

Soils can be considered an extension of a wick. The mobilization of any buried DDT in a soil to the atmosphere can be aided by the presence of water. DDT can be moved from deeper parts to the surface of a soil. There, if it precipitates from solution, it can vaporize more readily.

A second effect of water involves its ability to displace other molecules sorbed to a surface. DDT, or other halogenated hydrocarbons, can be strongly attached to dry soils or other dry surfaces. Upon wetting, the water molecules compete for sorption sites with the DDT. The strongly sorbed DDT molecules are less subject to vaporization than those displaced. This effect is called 'adsorption displacement' by Hartley.

These theoretical considerations can be extrapolated to predict DDT losses from agricultural areas. However, the complex breakdown patterns of DDT and the differing vapor pressures of its breakdown products, suggests that the relevant information will come from field studies.

Hexachlorobenzene (HCB)

Hexachlorobenzene, C_6Cl_6, the chlorinated homologue of benzene, is a powder with a vapour pressure of 1.1×10^{-5} mm of mercury at 20° C. It is used as a grain fungicide.

PRODUCTION

There are no production or sales statistics for the United States. However, the United States Department of Agriculture gives estimates of its use. Apparently 0.8 tons were used in 1966, with an increase to 6.9 tons in 1971. The sites of application are primarily in the State of Washington. Two United States concerns manufacture the chemical (Dover Manufacturing and Staufer). However, they are permitted by law to maintain the production and use data as privileged information.

There is widespread use of hexachlorobenzene (HCB) over the world. It is used extensively as a fungicide in Turkey, Italy, Spain, Netherlands, Federal Republic of Germany, France and some of the eastern European countries. Australia, alone, uses 180 tons of HCB in the treatment of wheat seed.

HCB is a contaminant in the pesticides diethyl tetrachloroterephthalate and pentachloronitrobenzene which may contain up to 10 per cent HCB. The pesticide pentachlorophenol may also contain HCB. It is a by-product in the manufacture of many chlorinated hydrocarbons such as perchloroethylene and carbon tetrachloride, which are produced in megaton amounts. Wastes are reported to contain 10–15 per cent HCB. If only 1 per cent of the production of low molecular weight halogenated hydrocarbon consisted of HCB, the annual production from just this source alone would be in the hundredths of megatons per year.

ENVIRONMENTAL FLUXES

The agricultural uses of HCB can be assumed to be totally dispersive. The vapour pressure, 100 times higher than that of DDT, suggests extensive evaporation of HCB into the atmosphere. The chemical wastes (containing HCB) if not incinerated, a rather uncommon practice, go into the ocean or into land fill. In known marine dumping areas, such as the Gulf of Mexico, the North Sea and the Norwegian Sea, there have been no reported measurements of HCBs in the waters. Vaporization from land fill is another potential route to the atmosphere and to the oceans.

ENVIRONMENTAL FATE

HCB is a very unreactive compound. There are no reports of photochemical degradation in the atmosphere or of hydrolysis in water. One possible metabolite, pentachlorobenzene, has been proposed (Stijve, 1971). It also appears quite refractory to breakdown by biological degradation processes.

ENVIRONMENTAL LEVELS

HCB is found in terrestrial and aquatic organisms. Measurements in marine organisms are collated in Table 12. The concentrations are generally lower than, but near, those of PCBs or DDT. This suggests similar fluxes and/or a greater environmental persistence.

TABLE 12. HCB concentrations in marine organisms

Species	Source	Number of specimens	Result		
			Mean	Units	Range
White perch	New Jersey, New York	3	0.22	μg/g wet weight	0.027–0.56
Striped bass (*Morone saxatilis*)	Maryland, Florida	5	0.038	μg/g wet weight	0.011–0.083
Chinook salmon eggs (*Oncorhynchus tshawytscha*)	Oregon	1	0.003	μg/g wet weight	—
Striped bass eggs	Maryland	8	0.23	μg/g wet weight	0–1.04
Striped bass	Florida	1	0.149	μg/g wet weight	—
Cod oil (*Gadus* sp.)	—	3	0.43	μg wet weight	0.100–0.63
Menhaden oil (*Brevoortia* sp.)	—	1	0.18	μg/g wet weight	—
American eel (*Anguilla rostrata*)	New Brunswick, St John river	13	0.012	μg/g wet weight	0.006–0.019
Eel liver			0.010	μg/g wet weight	—
Atlantic salmon (*Salmo salar.*)	New Brunswick	4	0.002	μg/g wet weight	—
Herring (*Clupea harengus*)	1 sample from Bay of Fundy and	10	0.006	μg/g wet weight	0.005–0.006
	2 from Chedabucto	10	0.004	μg/g wet weight	0.003–0.004
	Bay, Nova Scotia	10	0.003	μg/g wet weight	0.002–0.004
Mackerel (*Scomber scombrus*)	Bay of Fundy	4	0.001	μg/g wet weight	—
Seal (*Phoca vitulina*) Adult blubber	Coastal waters off the Netherlands	5	—	p.p.m., wet weight	<0.02–0.09
Juvenile blubber	Coastal waters off the Netherlands	5	0.067	p.p.m., wet weight	0.040–0.093
Bottle-nosed dolphin blubber (*Tursiops truncatus*)	Coastal waters off the Netherlands	2	—	p.p.m., wet weight	<0.013–0.032
Harbor porpoise blubber (*Phocoena phocoena*)	Coastal waters off the Netherlands	7	—	p.p.m., wet weight	<0.01–0.95

Adapted from NAS, *Assessing Potential Ocean Pollutants*, United States National Academy of Sciences, 1975, 438 p.

ENVIRONMENTAL IMPACTS

The consumption of wheat, coated with HCB and intended for planting, by inhabitants of south-eastern Turkey led to a porphyria-type disease. Victims who consumed roughly 50 to 200 mg HCB per day for relatively long time periods were affected. The total number of victims between 1955 and 1959 was estimated to be between 3,000 and 5,000.

Porphyria is a disturbance of the

metabolic pathway in heme synthesis which leads to an increase in the production of heme and its intermediates. Clinical manifestations include blistering and epidermolysis, abnormal pigmentation and arthritis.

Mirex

Mirex (dodecachlorooctahydro-1,3,4-methano-1H-cyclobuta(cd)pentalene) has been primarily used as an insecticide against the fire ant in the south-eastern United States, and secondarily as a flame retardent additive for polymer formulations. As a biocide, its employment in the United States has been concentrated in nine southern states, Florida, North Carolina, South Carolina, Mississippi, Louisiana, Texas, Alabama, Georgia, and Arkansas. By order of the United States Environmental Protection Agency in 1972, the emphasis of the fire-ant programme was shifted from a programme of eradication to one of control, and restrictions were placed upon its application.

The United States manufacturer is Hooker Chemical Company of Niagara Falls, New York. Production data are not available. However, its usage in the fire-ant programme gives a sense of United States output. Since 1967, more than 10 million acres per year have been treated in the fire-ant programmes. Normal applications involve 1.7 grammes of actual toxicant per acre (Markin *et al.*, 1974). Thus, for a single application per year, the lower limit to the estimated production would be 17 million tons. Multiple applications per year would raise the figure.

High levels of Mirex are reported to impact upon marine organisms. The larval stages of marine crabs were significantly affected by Mirex concentrations in the range of 0.01 to 10 p.p.b. in sea water with respect to survival. Also the duration of developmental stages and the total time of development is generally lengthened with an increase in Mirex concentration (Bookhout *et al.*, 1972). Concentrations as low as 0.1 p.p.b. may produce mortality in juvenile crayfish on the basis of laboratory experiments (Ludke *et al.*, 1971).

The heaviest infestations of the fire ant are in coastal regions and here one would expect leakage of Mirex to the marine environment. In 1971, seventy-seven composite samples of oysters, crabs, shrimp, fish and fish products were collected from the Gulf of Mexico and south-eastern United States coastal areas (Markin *et al.*, 1974). Only nine samples showed measureable Mirex (0.005–0.024 p.p.m.), all from near Savannah, Georgia. Clearly, Mirex is not as widespread a contaminant as the PCBs or DDT residues.

South Carolina coastal areas were sprayed aerially with Mirex fire-ant bait between October 1969 and December 1970. As a consequence, Mirex residues were detected in the following economically important members of the estuarine food chain: crabs, 0.60 p.p.m.; shrimps, 0–1.3 p.p.m.; and fishes, 0–0.82 p.p.m. Levels of Mirex in these organisms decreased to less than 0.01 p.p.m. over a period of eighteen to twenty-four months following the last aerial treatment (Borthwick *et al.*, 1974). Thus, the persistence of Mirex in these coastal waters, following the cut-off of the source, is low.

Finally, an unusual observation of Mirex in fish from Lake Ontario was reported by Kaiser (1974). Parts of fish from the Bay of Quinte had Mirex concentrations ranging between 0.020 and 0.050 p.p.m. The most probable source for this contamination is the manufacturing plant at Niagara Falls, located near Lake Ontario. Thus, an entry of this material from the production site to the Atlantic Ocean via the St Lawrence river is possible.

Low molecular weight halocarbons

The atmosphere and ocean contain a group of halocarbons with one or two carbon atoms which include dichlorodifluoromethane, trichlorofluoromethane, methyl iodide, trichloroethane, trichloroethylene, perchloroethylene, carbon tetrachloride and chloroform. The atmospheric concentrations of the chemicals are of the order of ng/m^3 and the surface ocean waters contain these chemicals in the range of ng/l. The fluorinated species of methane and the chloroethylenes are of human origin. The methyl iodide has been attributed to production by marine algae. There are no known marine sources for either chloroform or carbon tetrachloride; their concentrations in northern and southern hemispheric airs are similar, suggesting a natural source. Their origins have yet to be determined. Also, mixtures of short-chain aliphatic hydrocarbons, waste products from the manufacture of vinyl chloride, have been measured in the North Sea.

CHLOROFLUOROCARBONS

The two main members of this group are trichlorofluoromethane (CCl_3F—United States trade name, Freon 11) and dichlorodifluoromethane (CCl_2F_2—United States trade name, Freon 12). Both are primarily used as propellant solvents in aerosol dispensers with minor amounts employed as refrigerants. Since their propellant usage results in total dispersion to the atmosphere, their environmental fluxes are essentially the production values. For the United States, 0.12 and 0.18 million tons per year of trichlorofluoromethane and dichlorodifluoromethane were produced in 1968. A world production figure, three times this, or about a million tons per year has been proposed.

TRICHLOROETHANE

This chemical is primarily used as a cold cleaning solvent where there is a need to remove greases, oils or tars from surfaces at ambient temperatures. It is employed in the electronic industry, aircraft industry, textile industry, and petroleum industry.

It has a boiling point of 74º C at 760 mm of mercury and a water solubility of 4,400 p.p.m. by weight. United States production has increased from 9,000 to 169,000 tons per year over the period 1961 to 1971. A global production figure three times the 1971 value, or about 500,000 tons per year appears reasonable. Most of the above uses are dispersive, and thus their environmental fluxes are roughly equivalent to their rates of production.

TETRACHLOROETHYLENE

This compound, often referred to as Per, has major uses in dry-cleaning (65–70 per cent of United States production in 1972), vapour degreasing (17 per cent), the formation of chemical intermediates (11 per cent) and for other miscellaneous purposes (2–7 per cent). It is the principal solvent used in the dry-cleaning industry because it has excellent cleansing properties, is easily recycled, is non-flammable, and is not highly toxic.

It has a boiling point of 121º C at 760 mm of mercury pressure and a water solubility of 150 p.p.m. It has been produced commercially since before the First World War in Germany and the United Kingdom. In the 1920s and 1930s it replaced both trichloroethylene and carbon tetrachloride as dry-cleaning solvents. United States production has increased at an annual rate of 9 per cent over the past forty-five years and attained a value of 33,000 tons per year in 1972. Continued increases at this rate are contemplated for the foreseeable future. A world production of 1 million tons per year, three times the United States rate, is assumed. The uses are dispersive.

Environmental levels

Recent measurements of the chlorofluoro-carbons in marine and continental airs indicate a widespread and homogeneous distribution (except near densely populated areas in industrialized countries) as is shown below:

CCl_3F (Troposphere over North & South Pacific and ice-covered Ross Sea), $61\pm$ 13 p.p.t., Wilkness *et al.* (1973).

CCl_3F (San Diego, California, Troposphere), 287 ± 24, Su and Goldberg (1973).

CCl_2F_2 (San Diego, California, Troposphere), $3,150\pm142$, Su and Goldberg (1973).

CCl_3F (Troposphere over North and South Atlantic between the United Kingdom and Antarctica), 49.6 ± 7.1, Lovelock *et al.* (1973).

CCl_3F (Los Angeles basin, California, Troposphere), 650, Simmonds *et al.* (1974).

CCl_3F (Troposphere over Ireland), 10–190, Lovelock (1971).

$CHCl=CCl_2$ (Troposphere over United Kingdom), 11 ng/m³, Murray and Riley (1973).

$CCl_2=CCl_2$ (Troposphere over north-east Atlantic), 6; (Troposphere over United Kingdom), 19; (Troposphere over north-east Atlantic), 5; Murray and Riley (1973).

Several investigations have been made of their concentrations in surface sea water with the results as follows:

$CHCl=CCl_2$ (North-east Atlantic), 7 ng/l, Murray and Riley (1973).

$CCl_2=CCl_2$ (North-east Atlantic), 0.5 ng/l, Murray and Riley (1973).

CCl_3F (Atlantic between United Kingdom and Antarctica), 7.6 p.p.t., volume/volume, Lovelock *et al.* (1973).

CCl_2F_2 (La Jolla, California, Surface), 0.13 p.p.t., Su and Goldberg (1974).

CCl_3F (La Jolla, California, Surface), 0.041, Su and Goldberg (n.d.).

CH_3CCl_3 (La Jolla, California, Surface), 2.2, Su and Goldberg (n.d.).

$CHCl=CCl_2$ (La Jolla, California, Surface), 1.4, Su and Goldberg (n.d.).

$CCl_2=CCl_2$ (La Jolla, California, Surface), 1.4, Su and Goldberg (n.d.).

A model to ascertain the fluxes of such pollutant gases across the air/sea interface (Fig. 16) has been proposed by Liss and Slater (1974). A two-layer film system is put forth as the interface between air and water, both of which are assumed to be well mixed. The principal resistance to gas transport arises from the interfacial layers, where gas movement takes place by molecular processes. Using Fickian dynamics, and the sea-water and atmospheric CCl_3F values of Lovelock *et al.* (1973), Liss and Slater calculate an atmosphere to sea-water flux of CCl_3F over the entire oceans of 5.4×10^9 grammes per year. The postulated world production of CCl_3F is 0.3×10^{12} grammes per year. Since there are no known natural sources for this compound, it appears that about 2 per cent of the world's production is entering the oceans, and the rest remains in the atmosphere or is decomposed.

Environmental fates

Trichlorofluoromethane appears to be quite unreactive in the atmosphere. The chemical was irradiated for eight hours at room temperature and zero humidity with light wavelengths greater than 3,100 Å. (Japar *et al.*, 1974). The light intensities were comparable to those of Los Angeles at midday. There was no evidence of any reaction of this hydrocarbon or of the formation of any new products.

On the other hand, the chlorinated

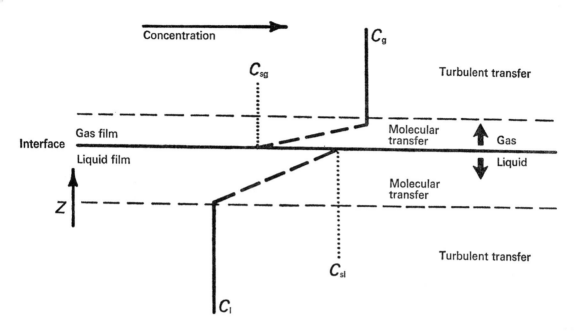

FIG. 16. The two-layer model of gas exchange processes in the ocean (Liss and Slater, 1974).

ethylenes and ethanes appear to be less stable in the atmosphere (NAS, 1975). The chlorinated olefins are readily oxidized by atmospheric oxygen. For example, perchloroethylene when irradiated with ultra-violet radiation in the presence of air or oxygen undergoes auto-oxidation to trichloroacetyl chloride.

Field studies have indicated that oxidants are produced from the exposure of the low molecular weight halogenated hydrocarbons to sunlight and that the amounts of oxidant formed are roughly inversely proportional to the stability of the compounds. The more stable chlorinated solvents such as perchloroethylene and trichloroethane produce far lesser amounts of oxidants than does the less-stable trichloroethylene. The latter disappears rapidly when exposed to sunlight. However, the Dow Chemical Company scientists found that perchloroethylene was more reactive than the trichloroethane when exposed to simulated sunlight under laboratory conditions. The present situation about the relative stability of the chlorinated compounds in the atmosphere is somewhat confused.

Hydrolysis, oxidation reactions and biological degradation can cause the disappearance of the hydrocarbons from sea water. Dow Chemical Company scientists have measured some of their half-lives in water under light and dark conditions (see Table 13).

Biological oxygen demands were not shown by trichloroethane or perchloroethylene for periods up to twenty days, thus

TABLE 13.

Compound	Half-life—dark (months)	Half-life—light (months)
$CCl_2=CCl_2$	9	6
CH_3CCl_3	6	6
$CHCl=CCl_2$	11	6

suggesting inorganic processes may be more significant in the environment. There is little work beyond this reported in the literature.

Although these compounds are not in steady state owing to increasing production of the chlorofluorocarbons residence times in the atmosphere can be estimated by assuming the steady-state situation. The average value of CCl_3F appears to be about 50 p.p.t. in the atmosphere away from centres of industrial and human activity. The volume of the atmosphere (corrected to STP) is 4.03×10^{21} l giving a CCl_3F content of 1.1×10^{12} g. Using a production rate of 0.36×10^{12} g per year, a residence time in the atmosphere of about three years is computed. Since the Liss and Slater air-sea flux is much smaller than the flux from a three-year residence time, it is apparent that much of the CCl_3F must be destroyed in the atmosphere or that there are other sinks, perhaps the polar ice sheets.

A similar calculation can be made for perchloroethylene using the measured value over the north-east Atlantic of 5 ng/m^3 as representative of the entire atmosphere. Thus, the atmospheric burden is 20×10^9 g. At an estimated world production rate of a million tons per year, and assuming all uses of perchloroethylene are dispersive, the residence time is computed to be seven days.

Aliphatic chlorinated hydrocarbons

Waste products from the production of vinyl chloride have been detected in the North Sea. The vinyl chloride is made from the chlorination of ethylene to produce initially 1,2-dichloroethane (EDC) which is subsequently converted to vinyl chloride. The vinyl chloride is polymerized to the polyvinyl chloride plastic PVC. The composition of the wastes, the so-called 'EDC tars' or 'bottoms' depends upon the impurities in the starting material ethylene and the conditions of manufacturing. A typical composition of the EDC tars is given in Table 14. Most of the EDC tar formed to date has been dumped into the oceans (Jernelöv, 1974).

The densities of the tars (1.25—1.40 g/cm^3) would suggest a tendency to sink in sea waters. However, its dispersion into smaller particles (perhaps colloidal in nature) and its tendency to adhere to a large variety of substances contribute to its retention in the upper layers of the ocean (Jernelöv, 1974).

The EDC tars have deleteriously affected some fish in the North Sea (Jensen *et al.*, 1970). Most fish specimens from the North Sea contain these components of EDC tars. Jernelöv reports that concentration factors for marine organisms range from 15 to 3,000. The biological half-life in cod liver varied between a few days for the low-boiling components to a few weeks for the less volatile ones.

Jernelöv estimates the European production of EDC tars to be 75,000 tons per year and a world production to be at least 300,000 tons per year. In the United States 77 per cent of the EDC produced in 1970 was consumed in the production of vinyl chloride and PVCs (Stanford Research Institute, 1972). Eighty companies in twenty-three other countries were listed as producing EDC in 1971 and there is evidence

TABLE 14. Composition by-products from vinyl chloride production using the oxychlorination process

By-product	Formula	Weight (%)
1,1,2-Trichloroethane	$CHCl_2—CH_2Cl$	67
1,2-Dichloroethane (EDC)	$CH_2Cl—CH_2Cl$	20
1,3-Dichlorobutane	$CH_3CHClCH_2CH_2Cl$	3
1,2-Dichlorobutane	$CH_3CH_2CHClCH_2Cl$	2
1,4-Dichloro-2-butene	$CH_2ClCH=CHCH_2Cl$	2
1,2,4-Trichloro-2-butene	$CH_2ClCH=CClCH_2Cl$	1.1
Pentachloroethane	$CHCl_2CCl_3$	1
Trichloroethylene	$CHCl=CCl_2$	1
asym. Tetrachloroethane	Cl_3CCH_2Cl	1
1,3-Dichloro-2-butene	$CH_3CCl=CHCH_2Cl$	0.5
1,2,4-Trichlorobutane	$CH_2ClCH_2CHClCH_2Cl$	0.3
bis (2-Chloroethylether)	$(CH_2ClCH_2)_2O$	0.3
1,3,4-Trichloro-1-butene	$CH_2ClCHClCH=CHCl$	0.3
1,1,2-Trichlorobutane	$CCl_2CHClCH_2CH_3$	0.3
2,3,4-Trichloro-1-butene	$CH_2ClCHClCCl=CH_2$	0.2

Adapted from A. S. Jensen, A. Jernelov, R. Lang and K. H. Palmork, 'Chlorinated Byproducts from Vinyl Chloride Production. A New Source of Marine Pollution', in *Proceedings of an FAO Technical Conference on Marine Pollution and its Effects on Living Resources and Fishing*, Rome, FAO, 1970.

that the rate of production is increasing with time.

Overview

The general environmental behaviours of the heavy halogenated hydrocarbons are now moderately well defined. The limited statistics available on production and uses of DDT and PCBs indicate that a sabbatical leave should not be taken from their surveillance in the marine environment. An examination of present knowledge may guide us in future actions both with regard to these substances and to other materials possessing similar characteristics that might challenge the resources of our surroundings. To gain an insight into what may be priority studies for the next few years with regard to the PCBs and DDT, we shall, first, summarize our present findings.

DDT, following application upon surfaces in agricultural and public-health activities, is partially degraded to DDE and other metabolites through biological or photochemical reactions. DDE, and perhaps other daughter compounds of DDT, have higher vapour pressures than their parent, and are preferentially evaporated to the atmosphere. The PCBs enter the atmosphere, and via this and other transport paths enter the oceans following dispersive usages in paints, plastics and paper products, and to a lesser extent following use in what are claimed to be closed systems such as large transformers and condensers. The DDT residues and the PCBs exist in the atmosphere mainly as gas molecules, not associated with particles. Following washout to the earth's surface, these substances enter the marine system. Part becomes associated with the biosphere, concentrating in lipid phases. The DDT residues are more easily degraded than the PCBs. The processes for their destruction are still undefined. There is a downward conveyance of both sets of compounds to deeper waters through their incorporation in living organisms and organic particles.

The biological impact of DDT residues has been moderately well established for the case of birds feeding upon marine organisms. Reproductive failures associated with egg-shell thinning have been documented in the case of pelicans. The halogenated hydrocarbons (DDT and PCBs) have also been implicated in the abortions of sea lion pups. No harmful effects of DDT upon human health have so far been observed. PCBs, on the other hand, if ingested, give rise to skin eruptions and a darkening of the epidermis, and in some cases may lead to deaths.

The low-molecular weight halocarbons are less well studied. Produced in megaton or fractions of a megaton quantities per year, they now have become ubiquitous components of the atmosphere and surface ocean waters. Although laboratory experiments have indicated that a high probability of atmospheric degradation is likely, the widespread finding of these species over the planet suggest they are more resistant to degradation than laboratory results indicate. Residence times in the atmosphere appear to be of the order of days.

With this background, three areas of concern have developed in my mind which I have characterized as 'the southward tilt', the 'global distillation', and the 'future of fermentation processes'.

Based upon informed estimates of DDT needs in cotton farming and in malarial control, there does not appear to be a possibility of a marked decline in production for the forthcoming decade. However, there is a geographical change in the use: the 'southward tilt' where increased applications will take place in the equatorial and southern hemispheric regions. As a consequence, surveillance programmes should pay greater attention to these areas. The 'southward tilt' in the use of DDT will probably not be paralleled in the use of PCBs. Present surveys indicate that its dispersive uses are decreasing and 'closed-system' applications are being carried out with greater safeguards against loss. The effects upon the ocean system of this shift in DDT use are not known.

The 'DDT story' has focused attention upon global processes. The extremely low vapour pressures of DDT and its residues sidetracked consideration of gas-phase transport by early investigators. However, the vast volume of the atmosphere, coupled with time scales of the order of years (far removed from the normal ones of days, in a laboratory) allowed the vaporization of vast amounts of DDT residues to take place. What other compounds of natural and/or anthropogenic sources are subject to the same process?

Natural or man-generated higher temperatures on the continents can initiate or increase the rate of distillation of organic substances to the oceans. For example, the burning of forests introduces substantial amounts of elemental carbon (soot) to the atmosphere and ultimately to the ocean sediments. This soot can constitute up to 0.1 per cent of the sedimentary solid phases. But what of other carbon-containing compounds that might be volatilized during this process? Dr Max Blumer of the Woods Hole Oceanographic Institution has indicated to me the widespread occurrence of condensed aromatics in marine sediments that might be attributed to such an origin. Some years ago, Dr C. Wendt proposed a distillation of plant volatiles, such as the terpenes, about the earth's surface. But I am even more concerned about what other synthetic organic chemicals there might be following the same route as DDT to the oceans.

What is called for is a systematic survey of organic chemicals in the atmosphere, in precipitation, in ocean waters and in marine sediments. The atmospheric and precipitation data will tell us the present fluxes; the sediment data on the basis of concentration changes as a function of depth in the deposits will indicate which resistant compounds may be natural and which may be man-made.

Finally, there is the spectre of the lower molecular weight halogenated hydrocarbons losing their apparent innocence as chemicals not impacting upon life processes. Professor John Wood of the University of Illinois has indicated that they may inhibit methyl transfer processes, such as occur in fermentation, through reactions with vitamin B–12. Their levels are building up in the environment. Laboratory investigations should be concerned with the levels at which fermentation, whether in sewage digestion or wine production, might be affected. Will these substances act additively? Is there any one or several among them that is especially effective along these lines?

These concerns convince me that there can be no let-up in investigations on the synthetic halogenated hydrocarbons, both heavy and light. Even if we stop the manufacture and use of some of these pollutants now, we cannot stop the continued entry of the already produced substances into the environment. Release will continue and some environmental levels can conceivably become higher before fall-off occurs.

Bibliography

ALBONE, E. S.; EGLINTON, G.; EVANS, N. C.; HUNTER, J. M.; RHEAD, M. M. 1972. Fate of DDT in Severn sediments. *Envir. Sci. Technol.*, vol. 6, p. 914–19.

BEVENUE, A.; HYLIN, J. W.; KAWANO, Y.; KELLEY, T. W. 1972. Organo pesticide residues in water, sediment, algae and fish, Hawaii—1970–1971. *Pestic. Monit. J.*, vol. 6, p. 60.

BIDLEMAN, T. F.; OLNEY, C. E. 1974. Chlorinated hydrocarbons in the Sargasso Sea atmosphere and surface water. *Science*, vol. 183, p. 516–18.

BLUS, L. J.; GISH, C. D.; BELISLE, A. A.; PROUTY, R. M. 1972. Logarithmic relationship of DDE residues to eggshell thinning. *Nature*, vol. 235, p. 376–7.

BOOKHOUT, C. G.; WILSON, A. J.; DUKE, T. W.; LOWE, J. I. 1972. Effects of mirex on the larval development of two crabs. *Water, Air, Soil Pollution*, vol. 1, p. 165–80.

BORTHWICK, P. W.; COOK, G. H., PATRICK Jr, J. M. 1974. Mirex residues in selected estuaries of South Carolina—June 1972. *Pestic. Monit. J.*, vol. 7, p. 144–5.

BUTLER, P. A. 1973. Residues in fish, wildlife and estuaries. *Pestic. Monit. J.*, vol. 6, p. 238–362.

CLIATH, M. M.; SPENCER, W. F. 1972. Dissipation of pesticides from soil by volatilization of degradation products. I: Lindane and DDT. *Envir. Sci. Technol.*, vol. 6, p. 910–14.

DELONG, R. L.; GILMARTIN, W. G.; SIMPSON, J. G. 1973. Premature births in California sea lions: association with high organochlorine pollutant residue levels. *Science*, vol. 181, p. 1168–9.

DUKE, T. W.; LOWE, J. I.; WILSON Jr, A. J. 1970. A polychlorinated biphenyl (Arochlor 1254) in the water, sediment and biota of Escambia Bay, Florida. *Bull. Environ. Contam. Toxicol.*, vol. 5, p. 171–80.

GIAM, C. S.; HANKS, A. R.; RICHARDSON, R. L.; SACKETT, W. M.; WONG, M. K. 1972. DDT, DDE and polychlorinated biphenyls in biota from the Gulf of Mexico and Caribbean Sea—1971. *Pestic. Monit. J.*, vol. 6, p. 139–43.

GIAM, C. S.; RICHARDSON, R. L.; WONG, M. K.; SACKETT, W. M. 1973. Polychlorinated biphenyls in antarctic biota. *Antarct. J.*, vol. 8, p. 303–5.

HALCROW, W.; MACKAY, O. W.; BOGAN, J. 1974. PCB levels in Clyde marine sediments and fauna. *Mar. Pollut. Bull.*, vol. 5, p. 134–6.

HARTLEY, G. S. 1969. Evaporization of pesticides. In: R. F. GOULD (ed.), *Pesticide formulations research*, p. 115–34. Washington, D.C., American Chemical Society.

HARVEY, G. R.; STEINHAUER, W. G.; MIKLAS, H. P. 1974*a*. Decline of PCB concentrations in North Atlantic surface water. *Nature*, vol. 252, p. 387–8.

HARVEY, G. R.; MIKLAS, H. P.; BOWEN, V. T.; STEINHAUER, W. G. 1974*b*. Observations on the distribution of chlorinated hydrocarbons in Atlantic Ocean organisms. *J. Mar. Res.*, vol. 32, p. 103–18.

HOLDEN, A. V. 1973. International co-operative study of organochlorine and mercury residues in wildlife, 1969–1971. *Pestic. Monit. J.*, vol. 7, p. 37–52.

HOM, W.; RISEBROUGH, R. W.; SOUTAR, A.; YOUNG, D. R. 1974. Deposition of DDE and polychlorinated biphenyls in dated sediments of the Santa Barbara Basin. *Science*, vol. 184, p. 1197–9.

IDOE. 1972. *Baseline studies of pollutants in the marine environment and research recommendations. The IDOE Baseline Conference, May 24–26, 1972 New York.* Washington, D.C., IDOE, National Science Foundation.

JAPAR, S.; PITTS, J. N.; WINER, A. M. n.d. *The photostability of fluorocarbons.* (Manuscript.)

JENSEN, S. A.; JERNELÖV, A.; LANG, R.; PALMORK, K. H. 1970. Chlorinated byproducts from vinyl chloride production. A new source of marine pollution. In: *Proceedings. FAO Technical Conference on Marine Pollution and its Effects on Living Resources and Fishing.* Rome, FAO, December 1970.

JENSEN, S.; SUNDSTROM, G. 1974. Structures and levels of most chlorobiphenyls in two technical PCB products and in human adipose tissue. *Ambio*, vol. 3, p. 70–6.

JERNELÖV, A. 1974. Heavy metals, metalloids and synthetic organics. In: E. D. GOLDBERG (ed.), *The Sea*, vol. 5, New York, Wiley Interscience.

KAISER, K. L. E. 1974. Mirex: an unrecognized contaminant of fishes from Lake Ontario. *Science*, vol. 185, p. 524–6.

KURATSUNE, M.; YOSIMURA, T.; MATSUZAKA, J.; YAMAGUCHI, A. 1972. Epidemiologic study on Yusho, a poisoning caused by ingestion of rice oil contaminated with a commercial brand of polychlorinated byphenyls. *Environmental health perspectives*, vol. 1, p. 119–28.

LISS, P. S.; SLATER, P. G. 1974. Flux of gases across the air-sea interface. *Nature*, vol. 247, p. 818.

LLOYD-JONES, C. P. 1971. The evaporation of DDT. *Nature*, vol. 229, p. 65–6.

LOVELOCK, J. E. 1971. Atmospheric fluorine compounds as indicators of air movements. *Nature*, vol. 230, p. 379.

LOVELOCK, J. E.; MAGGS, R. J.; WADE, R. J. 1973. Halogenated hydrocarbons in and over the Atlantic. *Nature*, vol. 241, p. 194–6.

LUDKE, J. L.; FINLEY, M. T.; LUSK, C. 1971. Toxicity of Mirex to crayfish, *Procambarus blandingi*. *Bull. Envir. Contam. Toxicol.*, vol. 6, p. 89–96.

MACKAY, D.; WOLKOFF, A. W. 1973. Rate of evaporation of low solubility contaminants from water bodies to atmosphere. *Environ. Sci. Technol.*, vol. 7, p. 611–14.

MARKIN, G. P.; HAWTHORNE, J. C.; COLLINS, H. L.; FORD, J. H. 1974. Levels of mirex and some other organochlorine residues in seafood from the Altantic and Gulf coastal states. *Pestic. Monit. J.*, vol. 7, p. 139–43.

MOSSER, J. L.; FISHER, N. S.; WURSTER, C. F. 1972. Polychlorinated biphenyls and DDT alterations of species composition in mixed culture of algae. *Science*, vol. 176, p. 533–5.

MURRAY, A. J.; RILEY, J. P. 1973. Occurrence of some chlorinated hydrocarbons in the environment. *Nature*, vol. 242, p. 37–8.

NAS. 1971. *Chlorinated hydrocarbons in the marine environment.* Washington, D.C., National Academy of Sciences. 42 p.

——. 1975. *Assessing potential ocean pollutants.* Washington, D.C., National Academy of Sciences. 438 p.

OECD. 1973. *Polychlorinated biphenyls, their use and control.* Paris. 44 p.

PATIL, K. C.; MATSUMARA, F.; BOUSH, C. M. 1972. Metabolic transformation of DDT, dieldrin, aldrin and endrin by marine microorganisms. *Environ. Sci. Technol.*, vol. 6, p. 629–32.

PIERCE, R. H.; OLNEY, C. E.; FELBECK, G. T. 1974. p,p'-DDT adsorption to suspended particulate matter in sea water. *Geochim. Cosmochim. Acta*, vol. 38, p. 1061–73.

RAFATJAH, H. A.; STILES, A. R. 1972. *Summary review of use and offtake of DDT in antimalaria control programs.* Geneva, World Health Organization. (VBC/72.5.)

RISEBROUGH, R. 1972. Cited in: Birds and pollution, an editorial article. *Nature*, vol. 240, p. 248.

SIMMONDS, P. G.; KERRIN, S. L.; LOVELOCK, J. E.; SHAIR, F. H. 1974. Distribution of atmospheric halocarbons in the air over the Los Angeles Basin. *Atmospheric environment*, vol. 8, p. 209–16.

SPENCER, W. F.; CLIATH, M. M. 1972. Volatility of DDT and related compounds. *J. Agr. Food Chem.*, vol. 20, p. 645–9.

STANFORD RESEARCH INSTITUTE. 1972. *Chemical economics handbook.* Palo Alto, Calif.

STANLEY, C. W.; BARNEY II., J. E.; HELTON, M. R.; YOBS, A. R. 1971. Measurement of atmospheric levels of pesticides. *Environ. Sci. Technol.*, vol. 5, p. 430–5.

STIJVE, T. 1971. Determination and occurrence of hexachlorobenzene residues. *Mitt. Geb. Lebensmittelunters. Hyg.*, vol. 62, p. 406–14.

SU, C.; GOLDBERG, E. D. 1973. Chlorofluorocarbons in the atmosphere. *Nature*, vol. 245, p. 27.

——. n.d. *Halocarbons in seawaters and the atmosphere.* (Manuscript.)

TARRANT, K. B.; TATTON, J. 1968. Organo-pesticides in rainwater in the British Isles. *Nature*, vol. 219, p. 725–7.

WARE, D. M.; ADDISON, R. F. 1973. PCB residues in plankton from the Gulf of St Lawrence. *Nature*, vol. 246, p. 519–21.

WEIL, L.; QUENTIN, K. E.; RONICKE, G. 1973. *Pestizidpegel des Luftstaubs in der Bundesrepublik.* Kommission zur Erforschung der Luftverunreinigung. Mitteilung VIII. 21 p.

WHITTEMORE, F. W. 1973. Personal communication. (Senior Officer, Plant Production Service, FAO, Rome.)

WILKNESS, P. E.; LAMONTAGNE, R. A.; LARSON, R. E.; SWINNERTON, J. W.; DICKSON, C. R.; THOMPSON, T. 1973. Atmospheric trace gases in the southern hemisphere. *Nature*, vol. 245, p. 45–7.

WILLIAMS, R.; HOLDEN, A. V. 1973. Organochlorine residues from plankton. *Mar. Poll. Bull.*, vol. 4, p. 109–11.

YATES, M. L.; HOLSWADE, W.; HIGER, A. L. 1970. *Pesticide residues in hydrobiological environments.* 159th ACS Natl. Meeting, Houston, Tex., Feb. 1970. Water, Air and Waste Chemistry Sec. of the Am. Chem. Soc. (Abstract p. Watr-032.)

4.   Radio-activity

The artificially produced radio-active species, arising from nuclear detonations or as waste products from nuclear plants, were the first group of substances recognized by the marine scientist as a potential challenge to the resources of the sea on a global basis. In principle, promiscuous release of these materials to the ocean system could result in widespread contamination. The mood of one of the first large-scale meetings of scientists concerned with the dangers of releasing radio-active materials to the oceans was summarized as follows (NAS-NRC, 1957):

The use of the sea for waste disposal, in particular, can jeopardize the other resources and hence should be done cautiously. . . . The large areas of uncertainty respecting the physical, chemical and biological processes lead to restrictions on what can now be regarded as safe practices. . . . If the sea is to be seriously considered as a dumping ground for any large fraction of the fission products that will be produced even within the next ten years, it is urgently necessary to learn about these processes to provide a basis for engineering estimates.

In addition, these scientists were aware that an understanding of the distributions of materials and predictions of future disseminations could only come about if an adequate book-keeping of environmental dispositions were available:

From the standpoint both of research and property control of this new kind of pollution careful records should be maintained of the kinds, quantities, and physical status of all radio isotopes introduced into the seas together with the detailed data on locations, depths and modes of introduction. This can probably best be done by national agencies reporting to an international record center.

Environmental scientists strongly recommended to authorities responsible for
the management of radio-active wastes to
minimize high-level releases to the marine
environment or to the atmosphere. Acceptable levels for the waters, plants and animals of the sea were developed on the basis
of potential dangers to man through his
ingestion of foods or through direct exposure. Discharges of nuclear wastes have
been regulated and dumping of high-level
radio-activities to the ocean has been pro-
hibited by the actions of individual nations.
The management of nuclear debris (excluding that released in weapons tests) has
become so effective, some scientists argue,
that the amount of radio-active material
reaching the marine environment per unit
of fuel burnt and reprocessed will probably
be reduced (Preston, 1972).

Other scientists, assessing the same information, indicate that such optimistic
views of our ability to protect marine resources from damage through release of
artificial radio-active materials may not be
warranted. Present concerns are directed
toward the transuranics to a somewhat
greater degree than towards the artificial
radionuclides produced as fission products
or as induced radio-active species, such as
zinc-65, cesium-137 and strontium-90, which
are accumulated in living organisms and
which can return to man in foods from the
sea. The transuranics, chemical elements
with atomic numbers greater than 92, are
produced in nuclear reactors and may subsequently be used as fuels or in nuclear
weapons. The members of this group, which
includes plutonium, neptunium, americium
and curium, are highly toxic. Bowen (1974)
points out that our present knowledge of
their biochemical and geochemical pathways in the marine environment is inadequate to develop the necessary risk-benefit
analyses to determine what quantities
would be acceptable in the oceans. The
amounts of the transuranics are continuing

to build up and it is predicted that by
the 1980s there will be hundreds of tons of
plutonium in nuclear reactors and in nuclear weapons. This contrasts with the
microgram quantities which can lead to
lung cancers if inhaled.

Before the oceans can be effectively and
routinely utilized as a receptacle for low-
level wastes from a large number of nuclear
installations, an extension of our present
information base is necessary. For example,
Bowen (*op. cit.*) points out that the pathways of the transuranics in the coastal zone
may differ for each geographical situation.
Thus for the siting of a single nuclear instal-
lation, those factors that may control the
environmental behaviours of the transuranics must be measured and used to
predict the ultimate fates of these toxic
materials. The accumulation of such knowledge will not come at a small cost.

Further, there is a great need to extend
our understanding of the interactions of
the radio-active species with living matter
(Templeton *et al.*, 1971). Up to the present,
regulatory actions upon the amounts of
radio-activity released to the oceans have
been based upon minimal jeopardies to
human health. As a consequence only a
modest effort has been made to understand
the impact of these substances upon individual organisms or upon communities of
organisms. Guides to laboratory and field
studies to consider such impacts might be
found in our knowledge of the specific en-
richments of chemicals by individual species
of organisms.

More formidable concerns may arise
from the development of a large number of
nuclear-power plants in the coastal areas of
the world. A large number of small leaks of
waste materials over long time periods can
eventually produce a dangerously radio-
active ocean.[1] Scientific resources for the

1. It should be recognized that the releases may not take
place directly from the power plants, and those from
the smaller number of reprocessing plants or those
occurring during the transport process are perhaps of
greater concern.

management of releases vary from one country to another, and the increasing need for energy sources in many nations may override the need for effective management policies. There is a lack of knowledge about what levels of each of the transuranics can be safely accommodated in the oceans, especially in coastal waters. Each local zone may have a unique set of pathways that could return transuranics, introduced into its waters, back to man's environment. The need for extended research efforts is emphasized in view of the growing possibility that nuclear-power plants will be sited offshore in the ocean. Present concepts involve floating rather than submerged facilities. The evaluation of potential offshore sites, based upon potential environmental effects, may stimulate additional work in radio-ecology.

Fortunately, several international agencies are now providing information relevant to the radio-active pollution of the oceans. Their outputs provide a basis on which nations can formulate policies concerning radio-active waste disposal in their coastal waters. The International Atomic Energy Agency assembles panels of experts to prepare reports on concepts and technologies in marine radio-active waste disposal. For example, analytical techniques are considered in *Reference Methods for Marine Radioactive Studies* (IAEA, 1970) and field investigations in *Radioactive Contamination of the Marine Environment* (IAEA, 1973). The International Commission on Radiological Protection (ICRP) provides compilations of acceptable body burdens in humans for various radionuclides. Such tabulations are used in developing monitoring strategies such as the 'critical pathways' approach (see Chapter 9).

Production

The greatest contribution of artificial radioactive nuclides to the ocean system has resulted from the atmospheric detonation of nuclear devices (Table 15) carried out by the United States, the United Kingdom, India, the U.S.S.R., the People's Republic of China and France. Up to 1968 there had been 470 nuclear explosions and the world oceans received a substantial part of the debris from them, except for those conducted underground or in outer space (Joseph *et al.*, 1971). The explosions are carried out for military purposes both in the air and underground, and for such civilian activities, usually subsurface, as excavation, mineral extraction and gas release.

TABLE 15. Total inventory of artificial radionuclides introduced into the world oceans

	1970	2000
Nuclear explosions *(world-wide distribution)*		
Fission products (exclusive of tritium)	$2-6 \times 10^8$ Ci	$? \times 10^8$ Ci*
Tritium	10^9 Ci	$? \times 10^9$ Ci*
Reactors and reprocessing of fuel (restricted local distribution)		
Fission and activation products (exclusive of tritium)	3×10^5 Ci	3×10^7 Ci
Tritium	3×10^5 Ci	$? \times 10^8$ Ci
Total artificial radio-activity	10^9 Ci	10^9 Ci
Total natural ^{40}K	5×10^{11} Ci	5×10^{11} Ci

* Assumed that atmospheric nuclear testing will continue at about the 1968–70 rate.
Reproduced from A. Preston, R. Fukai, H. L. Volchok and N. Yamagata, 'Report of the Panel on Radioactivity', in *Report of the Seminar on Methods of Detection, Measurement and Monitoring of Pollutants in the Marine Environment*, p. 87–99, Rome, FAO, 1971. (Marine Fisheries Reports, 99.)

The latter uses have not introduced substantial amounts of radio-active debris to the oceans. The amount of fissioned material has been estimated at 2.8×10^{28} fissioning atoms of uranium or plutonium. Two of the fission products which have entered the oceans with high activities, cesium-137 (half-life of thirty years) and strontium-90 (half-life of twenty-eight years) have been produced at levels of 21 and 34 megacuries, respectively (Joseph *et al.*, 1971). Of the explosions, 140 megatons or 72 per cent of the total yield, were produced as bursts that did not intersect the ground. Underwater detonations introduce radio-active species to rather localized areas initially, although subsequent mixing in the oceans can spread the nuclides over vast distances.

Land-based, and to a lesser extent ship-based, nuclear reactors for the production of steam or electricity generate significant amounts of radionuclides, but so far only a few cases have resulted in high marine concentrations.

Radio-isotopes are encapsulated for uses in energy production, in food preservation, in the sterilization of medical equipment, and in industrial thickness guages. Only in the case of rupture is there a possibility of oceanic entry. Three SNAP (Systems for Nuclear Auxiliary Power) reactors in satellites have left control of their managers. One was responsible for the entry of a significant part of the plutonium-238 in our atmosphere and oceans today. It should be noted that during the abortive operation these units behaved according to design.

Three types of artificially produced radio-active species are introduced to the oceans by man: (a) the nuclear fuels such as uranium-235 and plutonium-238; (b) the fission products, arising from the use of nuclear fuels, such as strontium-90 and cesium-137; and (c) the activation products, resulting from the interaction of nuclear particles with the components of nuclear reactors and weapons, such as zinc-65 and iron-55. The artificially produced radio-nuclides so far detected in the marine environment are given in Table 16. A wide spectrum of half-lives is noted for these nuclides, ranging from 1 day to 210,000 years.

The present unsettled situation over the future of nuclear-power plants makes indeterminable the number, types and locations of plants that will be in operation during the next few decades. Thus, accurate forecasts about the potential entry of radio-active wastes into the oceans cannot be made. For example, in September 1974 there were fifty-two operating nuclear plants in the United States with 187 plants under construction or on order. These planned plants would increase the energy output from nuclear installations sevenfold (Novick, 1974). However, somewhat after this time, it was announced that the construction of fifty-seven plants had been cancelled or delayed. Of this number, six or seven were already under construction. In addition, most of the active plants were operating at efficiencies much lower than initially planned. Unexpected technological and environmental problems were identified and at least for the time being have placed a damper on the combustion of nuclear fuels and the consequential production of waste products.

Herein, we will focus upon the potential production of the transuranics, as an example of the production of artificial radio-active nuclides. In addition, we will consider one of the unanticipated events following production, the accidental release of radioactivity to the ocean system through the abortion of a nuclear power satellite.

Transuranics are employed chiefly as fuels for the production of energy in reactors and in nuclear weapons. Plutonium-239 is the favoured nuclide because of its ready ability to capture neutrons and subsequently fission. The transuranics are also used as fuel in electrical and heat-generating units where their long replace-

TABLE 16. Artificially produced radionuclides which have been detected in the marine environment

	Mode of production				
	Fission			Neutron-activation	
Radio-nuclide	Half-life**	Type of decay	Radio-nuclide	Half-life	Type of decay
^{3}H*	12.26 y	β⁻	^{14}C*	5.76×10^3 y	β⁻
^{89}Sr	51 d	β⁻	^{32}P*	14.3 d	β⁻
^{90}Sr*	28 y	β⁻	^{35}S	87.2 d	β⁻
^{90}Y*	64.2 h	β⁻	^{45}Ca	165 d	β⁻
^{91}Y	59 d	β⁻	^{46}Sc	84 d	β⁻ γ
^{95}Nb*	35 d	β⁻ γ	^{51}Cr*	27.8 d	Kγ
^{95}Zr*	65 d	β⁻ γ	^{54}Mn*	314 d	Kγ
^{99}Mo	67 h	β⁻ γ	^{55}Fe*	2.7 y	K
^{99}Tc	2.1×10^5 y	β⁻	^{59}Fe	45 d	β⁻ γ
^{103}Ru*	40 d	β⁻ γ	^{57}Co	270 d	Kγ
^{106}Ru/^{106}Rh*	1.0 y	β⁻ γ	^{58}Co	71 d	Kβ⁺ γ
^{125}Sb	2.7 y	β⁻ γ	^{60}Co*	5.26 y	β⁻ γ
^{131}I	8.04 d	β⁻ γ	^{65}Zn*	245 d	Kβ⁺ γ
^{132}Te	78 h	β⁻ γ	^{76}As	26.5 h	β⁻ γ
^{137}Cs*	30 y	β⁻ γ	^{108m}Ag	127 y	Kγ
^{140}Ba	12.8 d	β⁻ γ	^{110m}Ag*	253 d	β⁻ γ
^{140}La	40.2 h	β⁻ γ	^{113m}Cd	14 y	β⁻
^{141}Ce	32.5 d	β⁻ γ	^{115m}Cd	43 d	β⁻ γ
^{144}Ce/^{144}Pr*	285 d	β⁻ γ	^{124}Sb	60 d	β⁻ γ
^{147}Pm	2.6 y	β⁻	^{134}Cs	2.1 y	β⁻ γ
^{155}Eu	1.81 d	β⁻ γ	^{181}W	30 d	Kγ
			^{185}W	73 d	β⁻
			^{187}W	24 h	β⁻
			^{207}Bi	28 y	Kγ
			^{239}Np	2.35 d	βγ
			^{238}Pu	86 y	α
			^{239}Pu*	2.44×10^4 y	α
			^{240}Pu	6.6×10^3 y	α
			^{241}Pu*	13.2 y	β
			^{241}Am*	458 y	αγ
			^{242}Cm	163 d	α

* Major recurring radionuclides.
** y, years; d, days; h, hours.
Reproduced from A. Preston, 'Artificial Radioactivity in Freshwater and Estuarine Systems', *Proc. Roy. Soc.* (London), Vol. 180B, p. 421–36.

ment times gives them an advantage over conventional materials. Plutonium-238 with its eighty-seven-year half-life is used extensively in the United States for such purposes. Today, the largest reservoir of plutonium-239 is in nuclear weapons, with an estimated mass of between 213,000 and 319,000 kilograms being deployed globally (NAS, 1975).

In the nuclear reactors, other transuranic nuclides will build up with time, the relative amounts depending upon the composition of the original fuel and the time spent in the reactor. Typical compositions of spent fuels are given in Table 17 for a light-water reactor (LWR) and for a liquid metal fast-breeder reactor (LMFBR) to illustrate the spectrum of nuclides produced. The nuclear fuels must be treated chemically every year or so to remove the unwanted products from nuclear reactions, the fission products, transuranic nuclides

TABLE 17. Examples of the transuranic content of spent nuclear fuels

Isotope	Light-water reactor	Liquid metal fast-breeder reactor
^{239}Np	0.05	2.0
^{238}Pu	8.5	3.2
^{239}Pu	1*	1*
^{240}Pu	1.5	1.2
^{241}Pu	350	170
^{241}Am	0.6	0.5
^{242}Cm	45	19
^{244}Cm	7.5	0.4

* Activities relative to that of plutonium-239.
Reproduced from NAS, *Assessing Potential Oceans Pollutants*, Washington, D.C., United States National Academy of Sciences, 438 p., 1975.

TABLE 18. Estimated annual quantities of plutonium in transit in the United States for 1980, 1990 and 2000 (in tons)

Category	1980	1990	2000
In shipment from inventory for fabrication	0	15	100
Spent fuel in shipment to reprocessing plant	7	33	160
Reprocessed wastes in shipment to storage	0	2	7

Source: NAS, *Assessing Potential Ocean Pollutants*, Washington, D.C., United States National Academy of Sciences, 438 p., 1975.

and induced activities in the inert parts of the fuel cells. The reprocessing plants recover the plutonium for continued use as a fuel (see Table 18).

By the year 1990, perhaps 50 tons of plutonium will be in a transport mode about the United States. Perhaps, double or triple this value will represent the movement of plutonium about the face of the earth. What will be the leakages to the environment? What amounts of this plutonium will enter the coastal environment? How long might it take for the development of pathways returning any accumulated plutonium in the coastal environment to man?

The management of plutonium constitutes one of the first-order problems in the development of nuclear technologies. Leakages from storage facilities and unintended releases have occurred. The unpredicted types of accidents that will occur are of special concern.

The accidental combustion of a navigational satellite containing a nuclear-power source has shown how a global dispersion of an atmospherically injected pollutant evolves and has provided basic knowledge about transport mechanisms of a transuranic at the earth's surface. The initial event occurred on 21 April 1964 when the SNAP-9A package did not reach orbital velocity because of a rocket failure after launch (Krey, 1967). The satellite apparently entered the atmosphere at a height of 46 kilometres over the Indian Ocean in the southern hemisphere. The device contained about 17 kilocuries of plutonium-238 and weighed about 12.3 kilograms. On the basis of the stratospheric inventory taken in 1966 (Krey, 1967) 88 per cent of the plutonium in the original source could be accounted for with 80 per cent in the southern hemisphere stratosphere while 20 per cent was transported to the northern hemisphere. By the end of 1970, it was estimated that less than 1 kilocurie remained

TABLE 19. Inventory of plutonium-239, 240 and plutonium-238 fallout (in kilocuries)

Area	kCi deposited		
	Plutonium-239, 240	Plutonium-238	
		Weapons	SNAP-9A
Northern hemisphere	256 ± 33	6.1 ± 0.8	3.1 ± 0.8
Southern hemisphere	69 ± 14	1.6 ± 0.3	10.8 ± 2.1
Global	325 ± 36	7.7 ± 0.9	13.9 ± 2.2

Adapted from E. P. Hardy, P. W. Krey and H. L. Volchok, 'Global Inventory and Distribution of Fallout Plutonium', *Nature*, Vol. 241, p. 444–5, 1973.

above 12 kilometres and by mid-1970, 95 per cent of the plutonium-238 was deposited on the earth's surface (Hardy *et al.*, 1973). The amount of plutonium-238 which had been introduced to the environment in weapons testing was doubled by that from the SNAP device (Table 19).

In 1970, soil profiles indicated that there were measurable plutonium-238 concentrations from the SNAP event up to 70° N. latitude down to 44° S. latitude (Hardy *et al.*, 1973). Of importance is the observation that within six years after the stratospheric injection, nearly all of the plutonium had fallen to earth over an area that included the highly populated zones of the mid-latitudes of the northern hemisphere. The stratosphere can be a distributor of pollutants about the surface of the earth.

Sea-water inventories

The total sea-water burden of artificial radio-activities is less than 1 per cent of the natural level in activity units (Table 15). The contributions from nuclear explosions in 1970 dominated those from reactor wastes and from the reprocessing of fuels; by the year 2000 they will still be an order of magnitude higher, assuming atmospheric nuclear testing will continue to take place at the 1968–70 rate. However, the improved techniques in radio-active waste management may reduce these predicted values substantially (Preston *et al.*, 1971).

The artificial radio-activities will not be distributed as homogeneously within the ocean system as the two most important naturally occurring radionuclides, potassium-40 and rubidium-87.[1] This can be seen very well from Table 20 where the levels of the major fallout radionuclides in surface sea waters have been collated (Woodhead, 1973). Higher values of all fallout species are found in the northern as compared to the southern hemisphere. The greater number of nuclear-bomb detonations took place there. Similarly, the higher values of radio-active wastes associated with the production and use of nuclear fuels will be found in marine waters near the waste discharge sites. In such cases the levels of a given isotope are far in excess of the open sea values where entry has been primarily from atmospheric fallout.

The distribution of a radionuclide emitted from a power or reprocessing plant, whose stable counterparts have conservative properties in the ocean system, can well assist in studies of coastal circulation. Such has been the case with the cesium-137

1. With respect to a natural background dose rate to all groups of organisms, alpha-emitting nuclides, especially polonium-210, are the most significant (Woodhead, 1973).

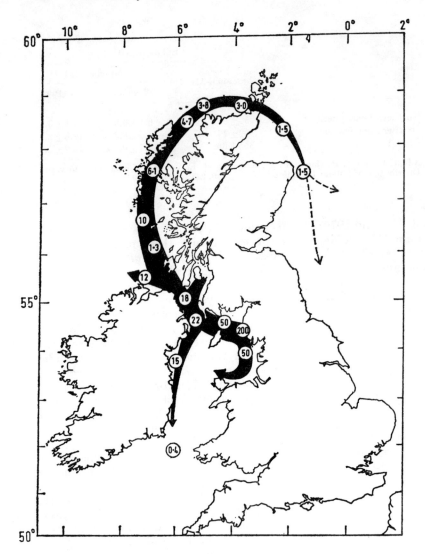

FIG. 17. Concentration of cesium-137 (picocuries/litre) in British Isles coastal water for May-July 1972 (Jeffries *et al.*, 1973).

released in the north-east Irish Sea from the Windscale nuclear fuel processing plant (Jeffries *et al.*, 1973). Monitoring surveys provided the data that indicate the water tagged by the cesium-137 in the vicinity of the discharge site travels northward, hugging the coasts. Moving clockwise around the west Scottish coast, it enters the northern North Sea where it penetrates in measurable amount to the northern coasts of England. Of special significance from this study is the observation of how well the isotope is maintained in the coastal waters following release (Fig. 17).

Table 20. Levels of the major fallout radionuclides in surface sea water

Location	Average concentration and/or range (pCi/l)				
	^{90}Sr	^{137}Cs*	^{3}H	^{14}C	^{239}Pu
North Atlantic Ocean	0.13 (0.02–0.50)	0.21 (0.03–0.8)	48 (31–74)	0.02 (0.01–0.04)	$(0.3–1.2) \times 10^{-3}$
South Atlantic Ocean	0.07 (0.02–0.20)	0.11 (0.03–0.32)	19 (16–22)	0.03 (0.02–0.04)	0.2×10^{-3}
Indian Ocean	0.10 (0.02–0.15)	0.16 (0.03–0.24)	—	—	—
North-West Pacific Ocean	0.54 (0.07–3.1)	0.86 (0.11–5.0)	29 (6–70)	0.03 (0.02–0.03)	$(0.1–1.4) \times 10^{-3}$
South-West Pacific Ocean	0.08 (0.01–0.20)	0.13 (0.02–0.32)	8 (0.7–22)	—	—
North-East Pacific Ocean	0.27 (0.05–0.58)	0.43 (0.08–0.93)	44 (10–240)	0.03 (0–0.04)	$(0.1–1.3) \times 10^{-3}$
South-East Pacific Ocean	0.09 (0.03–0.33)	0.14 (0.05–0.53)	8 (0.3–34)	0.01 (0–0.03)	—
North Sea	0.50 (0.31–0.97)	0.80 (0.50–1.55)	—	—	—
Baltic Sea	0.71 (0.36–1.0)	1.1 (0.56–1.6)	—	—	—
Black Sea	0.47 (0.07–0.78)	0.75 (0.11–1.25)	—	—	—
Mediterranean Sea	0.23 (0.09–0.38)	0.37 (0.14–0.61)	—	—	—

* Calculated from the ^{90}Sr values on the assumption that the activity ratio ^{137}Cs/^{90}Sr $=1.6$.
Reproduced from D. S. Woodhead, 'Levels of Radioactivity in the Marine Environment and the Dose Commitment to Marine Organisms', in *Radioactive Contamination of the Marine Environment*, p. 499–525, Vienna, International Atomic Energy Agency, 1973.

Strontium-90 and cesium-137

Strontium-90 and cesium-137 are two of the more intensively studied artificially produced isotopes in the ocean system. They are considered to be transported solely with the water and hence are often used as tracers of water masses (Volchok *et al.*, 1971). In the Atlantic, strontium-90 decreases more or less regularly with depth down to 700 metres. The maximum is at the surface and concentrations rarely fall below 10 per cent of the surface value (Fig. 18). On the other hand, in the north-west Pacific, the 700 metres values are about 35 per cent of the surface values. In the north-east Pacific, penetrations below 700 metres depth did not occur. In these waters the surfaces values were maintained to about 150 metres after which there was a very rapid falloff to 700 metres. In California coastal waters the falloff is even more rapid with practically no strontium-90 found below 400 metres.

There has been disagreement as to the occurrence of strontium-90 and cesium-137 in waters deeper than 1,000 metres (Volchok *et al.*, 1971). The arguments concerning the validity of data from various laboratories revolve around the measurement of reagent blanks. Preston (1974) points out that the oceanic inventory of strontium-90 would appear to be much larger than that estimated from known nuclear-test explosions and that vertical mixing rates would need to be much higher than generally accepted, if strontium-90 has penetrated to such depths.

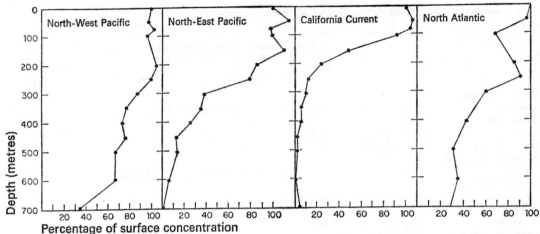

FIG. 18. The strontium-90 concentrations in the Atlantic and in the Pacific Ocean (Volchok *et al.*, 1971).

Carbon-14 and tritium

Two artificially produced radionuclides, hydrogen-3 (tritium) and carbon-14, have naturally occurring counterparts produced through the cosmic-ray fragmentation of atmospheric gases. The global inventory of tritium now contains greater contributions from the detonation of megaton-fusion bombs than from its natural production (Volchok *et al.*, 1971). There was a hundred-fold increase in the tritium level as a result of hydrogen-bomb testing in 1962 and 1963. Nearly all of the tritium, regardless of its mode of formation, is in the form of tritiated water (Martell, 1963). This nuclide thus becomes a potential tracer of the surface-water mass into which it is introduced. With its 12.26 year half-life, it can be used to study circulation processes over time periods of about fifty years.

The pulses of tritium that enter the ocean from stratospheric fallout in late winter and early spring are rapidly accommodated throughout the mixed layer (Dockins *et al.*, 1967) in the North Pacific.

In the northern latitudes of the Atlantic there is penetration of tritium through the thermocline (Roether and Munnich, 1967) whereas in equatorial waters it is maintained within the mixed layer.

Before 1950 the world-wide inventory of carbon-14 was 2.2×10^{30} atoms; this has been increased through the detonation of nuclear devices by 6×10^{28} atoms (Fairhall and Young, 1970). Penetrations of carbon-14 to depths of nearly 1,000 metres were noted by these workers. In the North Pacific it appeared that more than two-thirds of the anthropogenic carbon-14 was below the mixed layer. Lower activities of carbon-14 are found in upwelling areas since deeper, low carbon-14 waters dilute the bomb-produced carbon-14 activity in the surface waters (Bien and Suess, 1967).

OTHER FISSION PRODUCTS

A group of fission products, zirconium-95, niobium-95, ruthenium-103, ruthenium-106, cerium-141, cerium-144 and promethium-147, after entry to the marine environment, become associated with particulate phases (Volchok *et al.*, 1971). Such nuclides show irregular depth profiles, sometimes with secondary maxima in deep waters. Whether these particles have sunk with the debris of atmospheric weapons tests or whether their conveyance to deeper waters primarily involves the biosphere or the sinking of inorganic particles is still unresolved. In general, these radionuclides do show similar concentration/depth profiles (Preston, 1974). This suggests that sedimentation may be the controlling factor, inasmuch as interactions with the biosphere would most probably produce marked and different fractionations between the water and biosphere for each radionuclide.

Plutonium

Most of the plutonium in the ocean system today came as a result of nuclear weapons tests and the failure of a plutonium-238 power source. Plutonium-239 activities in sea water are only a few per cent of the level of activity of strontium-90 and cesium-137, while plutonium-238 is only a few per cent of that of the isotope plutonium-239. The plutonium residence time in the upper layers of the ocean is about four years in contrast to that of ten years for cesium (T. Folsom, personal communication). Plutonium in sea water is highly reactive, unlike cesium which is only moderately involved in chemical reactions in the oceans. Plutonium is rapidly taken up on the surfaces of large algae. Its chemistry appears similar to that of polonium (Folsom, personal communication). Like polonium its conveyance out of surface waters probably results from its tendency to adhere to both organic and inorganic surfaces.

Analyses of soils from undisturbed areas have given a measure of the integrated fallout of plutonium isotopes to the earth's surface (Hardy *et al.*, 1973). The plutonium

was recovered from ten 8.9 cm diameter cores taken to 30 cm depth, giving a sampling area of 622 cm². The plutonium-239 and 240 isotopes were assumed to have derived completely from weapons testing, whereas the plutonium-238 was derived from that source as well as from the SNAP-9A burn-up in 1964. Sampling was done in sixty-five sites around the world. The plutonium isotopes from this large sample were removed by an acid leaching which was found to be as effective as total dissolution of the sample.

The latitudinal distribution of the fallout is given in Table 21. The weapons component of the total plutonium-238 fallout is obtained by multiplying the amount of plutonium-239, 240 by 0.024, the average plutonium-238/plutonium-239, 240 ratio before the SNAP-9A accident. The plutonium fallout from weapons is greatest in the mid-latitudes of the northern hemisphere, with a minimum occurring in the equatorial regions. The maximum in the southern hemisphere is about one-fifth of that of the northern hemisphere. The introduction of the plutonium took place predominantly in the northern hemisphere, and the weapons fallout clearly reflects some stratospheric mixing.

The SNAP-9A plutonium has a two and a half times greater deposition in the southern hemisphere than in the northern in accordance with the unintended re-entry and burn-up of the plutonium-238 energy source in the southern hemisphere. The total fallout and that for each of the hemispheres is given in Table 19, based upon extrapolation of the soil sample data to the oceans.

However, limited studies on the integrated deposition of plutonium in the marine environment indicates a greater fallout in the ocean than on land at a similar latitude (Koide *et al.*, 1975). The following model is proposed to account for this. Whereas the plutonium that enters the

TABLE 21. Average latitudinal distributions of cumulative plutonium-239, 240 and plutonium-238 fallout

Hemisphere	Latitude band⁰	mCi per km²		
		Plutonium-239, 240	Plutonium-238	
			Weapons	SNAP-9A
Northern	90–80	(0.10 ± 0.04)	(0.002 ± 0.001)	(<0.001)
	80–70	0.36 ± 0.05	0.009 ± 0.001	<0.001
	70–60	1.6 ± 1.0	0.038 ± 0.025	0.026 ± 0.015
	60–50	1.3 ± 0.2	0.031 ± 0.004	0.013 ± 0.004
	50–40	2.2 ± 0.5	0.053 ± 0.011	0.026 ± 0.011
	40–30	1.8 ± 0.6	0.042 ± 0.014	0.025 ± 0.015
	30–20	0.96 ± 0.07	0.023 ± 0.002	0.011 ± 0.004
	20–10	0.24 ± 0.10	0.006 ± 0.002	0.003 ± 0.002
	10–0	0.13 ± 0.06	0.003 ± 0.001	<0.001
Southern	0–10	0.30 ± 0.20	0.007 ± 0.005	0.010 ± 0.007
	10–20	0.18 ± 0.05	0.004 ± 0.001	0.036 ± 0.021
	20–30	0.39 ± 0.16	0.009 ± 0.004	0.070 ± 0.042
	30–40	0.40 ± 0.12	0.009 ± 0.003	0.061 ± 0.020
	40–50	0.35 ± 0.21	0.008 ± 0.005	0.069 ± 0.038
	50–60	(0.20 ± 0.09)	(0.005 ± 0.002)	(0.044 ± 0.023)
	60–70	(0.10 ± 0.04)	(0.002 ± 0.001)	(0.022 ± 0.012)
	70–80	(0.03 ± 0.01)	(0.001 ± 0.001)	(0.008 ± 0.005)
	80–90	(0.01 ± 0.004)	(<0.001)	(0.004 ± 0.002)

Note: results in parentheses were derived by extrapolation.
Reproduced from E. P. Hardy, P. W. Krey and H. L. Volchok, 'Global Inventory and Distribution of Fallout Plutonium', *Nature*, Vol. 241, 1973, p. 444–5.

TABLE 22. Plutonium isotopes in atmospheric dusts and in recent marine sediments

Sample	Plutonium-239, 240 (dpm/g)	Plutonium-238 (dpm/g)
Atmospheric dust from Santa Barbara, California, 1974	2.8	0.11
Santa Barbara basin sediment, strata from 1973–74	1.0	0.025
Atmospheric dust from La Jolla, California, 1974	2.2	0.064
Soledad basin sediment, Baja California, 1973	1.0	0.028

Adapted from M. Koide, J. J. Griffin and E. D. Goldberg, 'Records of Plutonium Fallout in Marine and Terrestrial Samples', *J. Geophys. Res.*, Vol. 80, 1975, p. 4153–62.

ocean system probably is rapidly transferred to the sediments, and becomes locked into the deposit site, that which is accommodated in soils can be subsequently mobilized by winds or rivers. In part, this remobilized plutonium is transported to the marine environment where it becomes incorporated in the sediments.

Field observations made in the western United States are in accordance with this model (Koide *et al.*, 1975). The plutonium contents of airborne dust are similar to those of the surface coastal sediments (Table 22). Whereas a maximum in plutonium concentrations would be expected in strata deposited around 1963, the year of

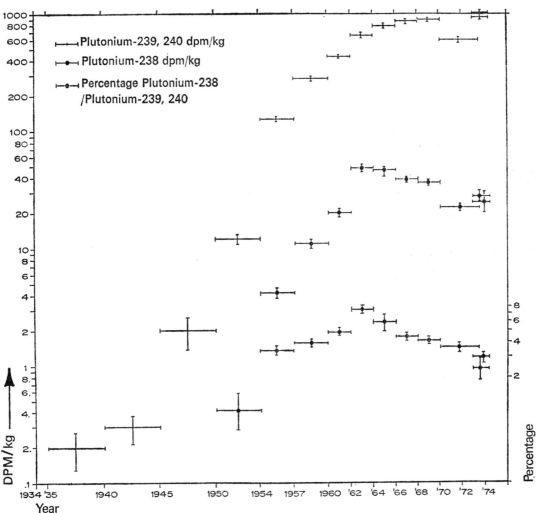

FIG. 19. Plutonium isotopes in yearly strata of the Santa Barbara basin deposits (Koide *et al.*, 1975. Copyright by the American Geophysical Union).

the highest fallout, there is an increasing flux of plutonium to these deposits up to the 1970s where it may be levelling off (Fig. 19). No mobilization of the plutonium within the deposits is indicated on the basis of the sharp cut-off between the 1945–50 and 1950–54 levels, the break between the period of no bomb tests and the initiation of extensive testing. The most reasonable explanation for this apparent anomaly is an additional source of plutonium to the deposits other than stratospheric fallout.

The integrated fallout of plutonium-239, 240 into the Santa Barbara basin is 4.5 mCi/km² which is based upon an accumulation rate of 0.4 cm per year and a sediment density of 0.225 g/cm³. This may be compared to the average value of 1.8 mCi/km² for 30–40º N. latitude obtained by Hardy *et al.* (1973) based on soil profiles and that of Noshkin (1972) of 2.3 mCi/km² for Buzzards Bay (41º N. and 70º W.). The Santa Barbara basin flux is substantially higher and permissively supports the contention that there is an additional source of plutonium isotopes, besides stratospheric fallout, to the deposit site. In addition, the mineralogy and size distributions of the dust particles are similar to those found in the Santa Barbara basin deposits.

This remobilization of stratospheric fallout may explain one of the persistent enigmas that has confronted the bookkeepers of artificial radio-activity on earth: the strontium-90 fallout over the sea appears to be double that over land (Volchok *et al.*, 1971). Volchok *et al.* indicate that there is 'no reasonable explanation for this inconsistency'.

These plutonium results provide an explanation for the strontium problem if there is a similar transfer of strontium-90 fallout accommodated in soils and weathered rock debris from the continents to the oceans.

A marked enrichment in marine organisms over sea water has been noted

(Noshkin *et al.*, 1972). The concentration factors of plutonium-239 for marine invertebrates, the ratio between the activity (dpm) per unit mass fresh (wet weight) organism and the concentration in the same mass of average North Atlantic sea water (taken as 0.19 dpm/100 kg) is shown in Table 23.

The blue mussel, *Mytilus edulis*, has similar concentrations of plutonium and values of the plutonium-238/plutonium-239 ratio in both the body and shell (Table 23). Noshkin *et al.* suggest that this indicates that plutonium is rapidly exchanged across the mantle, the site involved in both food collection and shell formation. Further, they propose that *Mytilus* would be a most useful biological indicator for the environmental levels of plutonium. These organisms, attached to piles, rocks, etc., are usually removed from direct contact with bottom detritus and probably receive their plutonium from the food they consume.

The higher plutonium concentrations appear to be in organisms feeding on sedimentary particles or upon surfaces. The marine worm (Table 23) had the highest plutonium content of the invertebrates examined. This species, *Nereis*, is a nonselective bottom feeder which ingests significant quantities of sediment. Noshkin *et al.* note that the surface sediments have higher plutonium concentrations than do the overlying waters. The worm was dredged from a scallop bed. Noshkin *et al.* suggest that the sponge acquires its plutonium both from occluded detritus and burrowing worms that use the sponge as a host.

The sea-weed, *Sargassum*, appears to be the most efficient collector of plutonium so far examined. The standing crop of Sargasso weed in the Sargasso Sea is about 1.5×10^3 kg/km² and the area of the sea is 5.2×10^6 km². Taking an average value of 46 dpm/kg for plutonium-239 there is a standing crop of 0.2 curies of plutonium-239

TABLE 23. The plutonium isotopic levels in marine organisms

Marine invertebrates	Collection date	Tissue Analysed	Tissue Wet weight (kg)	^{239}Pu dpm/100 kg (wet weight)	$\dfrac{^{238}Pu}{^{239}Pu}$
Blue mussel (*Mytilus edulis*)					
Cape Cod Canal, east end	VI.70	Body	1.76	51 ± 19**	—
Cape Cod Canal, east end	VI.70	Body	0.94	56 ± 8**	0.099
Woodneck Beach	III.70	Body	0.37	62 ± 11**	—
Woodneck Beach	VI.70	Body	0.54	56 ± 8**	—
Dyers Dock	VII.70	Body	0.75	36 ± 5**	—
Millstone Beach, Conn.	XI.70	Body	0.63	97 ± 11	—
Plymouth Harbor	III.71	Body	0.43	55 ± 8	—
Woodneck Beach	III.70	Shell	0.48	89 ± 14**	—
Millstone Beach, Conn.	XI.70	Shell	0.78	98 ± 8	0.081
Cape Cod Canal, west end	XI.70	Shell	0.51	92 ± 14	0.095
Brown mussel (*Modiolus modiolus*)					
Hadley Island	VI.70	Body	0.97	64 ± 6**	0.071
Softshell clam (*Mya arenaria*)					
Hadley Island	VI.70	Body	0.28	83 ± 17**	—
Oyster (*Ostrea virginica*)					
Waquoit Bay	X.70	Body	0.52	19 ± 5	—
Waquoit Bay	X.70	Body	0.51	31 ± 5	—
Scallop (*Pecten irradians*)					
Osterville	X.70	a.m.*	0.53	2 ± 1	—
Waquoit	X.70	a.m.	0.61	7 ± 2	—
Tobey Island	X.70	Body	1.04	131 ± 6	0.041
Osterville	X.70	Body	0.63	87 ± 8	0.057
Waquoit	X.70	Body	0.66	78 ± 8	—
Waquoit	X.70	Shell	0.51	115 ± 11	0.058
Whelk (*Busycon corira*)					
Waquoit	X.70	Body	1.81	27 ± 2	—
Waquoit	X.70	Shell	0.43	56 ± 6	0.093
Moon shell (*Lunatia heros*)					
Manensha	X.70	Body	0.42	126 ± 9	0.046
Manensha	X.70	Shell	0.46	131 ± 9	—
Starfish (*Asterias forbesi*)					
Cape Cod Canal, east end	VI.70	Body	1.40	220 ± 8**	0.085
Dyers Dock	VI.70	Body	1.77	167 ± 20	0.090
Brittle star (*Ophiuroidea*)					
Waquoit	XI.70	Body	0.54	145 ± 15	0.090
Marine worm					
Barnstable	X.70	Body	0.61	778 ± 42	0.056
Sponge (*Clathria delicata*)					
Waquoit	XI.70	Body	0.41	399 ± 51	0.074

(*continued*)

* a.m., adductor muscle.
** Corrected data from K. M. Wong, J. C. Burke and V. T. Bowen, 'Plutonium Concentrations in Organisms of the Atlantic Ocean', *Health Physics Society Annual Symposium: Proceedings of Midyear Topical Symposium*, p. 529–39.

TABLE 23. The plutonium isotopic levels in marine organisms (*continued*)

| Marine invertebrates | Collection date | Tissue | | ^{239}Pu dpm/100 kg (wet weight) | $\dfrac{^{238}Pu}{^{239}Pu}$ |
		Analysed	Wet weight (kg)		
Phytoplankton					
Sargasso weed (*Sargassum sp.*)					
25º N., 60º W.	IV.70		0.32	1,990 ± 75*	0.051
40º N., 54º W.	VII.69		0.76	911 ± 35	0.092
25º N., 67º W.	XI.68		0.22	624 ± 30	0.026
22º N., 65º W.	V.66		0.25**	1,070 ± 47	0.030
28º N., 68º W.	II.66		0.05**	18,500 ± 62	0.035
22º N., 48º W.	XII.65		0.14**	4,500 ± 32	0.057
Staghorn (*Codium fragile*)					
Dyers Dock	XI.70		0.64	39 ± 7	—
Fucus vessaculosis					
Dyers Dock	II.71		0.41	139 ± 22	—
Ascophyllum nodisum					
Plymouth Harbor	III.71		0.51	126 ± 13	—
Plymouth Harbor	III.71		0.51	301 ± 20	—
Chondrus crispus					
Plymouth Harbor			0.39	76 ± 37	—
Kelp					
Quisset Beach	III.71		0.47	20 ± 8	—

* Corrected data from K. M. Wong, J. C. Burke and V. T. Bowen, 'Plutonium Concentrations in Organisms of the Atlantic Ocean', *Health Physics Society Annual Symposium: Proceedings of Midyear Topical Symposium*, p. 529–39.
** Estimated wet weight from dry weight.
Adapted from V. E. Noshkin, V. T. Bowen, K. M. Wong and J. C. Burke, 'Plutonium in North Atlantic Ocean Organisms: Ecological Relationships', in D. J. Nelson (ed.), *Radionuclides in Ecosystems: Proc. Third Nat. Symp. Radioecology, 10–12 May 1971, Oakridge, Tenn.*, Vol. 2, p. 681–8.

in the sea-weed. Since the crop is regenerated annually, this amount of plutonium can be transported to the marginal zones of the ocean or to the seashore where it can die and decompose. Such a mobilization could result in a concentration of plutonium in the coastal zone.

Wong *et al.* (1972) also found high concentrations of plutonium (as well as polonium) in marine algae. In the giant brown eel kelp, *Pelagophycus porra*, highest levels were found on the outermost parts of the plant; such areas may be most sensitive indicators of environmental levels.

Bibliography

BIEN, G. S.; SUESS, H. E. 1967. Transfer and exchange of C-14 between the atmosphere and surface water of the Pacific Ocean. In: *Proceedings of the symposium on radioactivity dating and methods of low level counting.* Vienna, International Atomic Energy Agency, p. 105–15.

BOWEN, V. T. 1974. Transuranic elements and nuclear wastes. *Oceanus*, vol. 18, p. 43–54.

DOCKINS, K. O.; BAINBRIDGE, A. E.; HOUTERMANS, J. C.; SUESS, H. E. 1967. Tritium in the mixed layer of the North Pacific Ocean. In: *Proceedings of the symposium on radioactivity dating and methods of low level counting.* Vienna, International Atomic Energy Agency, p. 129–41.

FAIRHALL, A. W.; YOUNG, J. A. 1970. Radiocarbon in the environment. In: *Radionuclides in the environment,* p. 401–18. (Advances in chemistry series, no. 93.) American Chemical Society.

HARDY, E. P.; KREY, P. W.; VOLCHOK, H. L. 1973. Global inventory and distribution of fallout plutonium. *Nature,* vol. 241, p. 444–5.

IAEA. 1970. *Reference methods for marine radioactive studies.* Vienna, 284 p. (International Atomic Energy Agency technical report series, No. 118.)

——. 1973. *Radioactive contamination of the marine environment.* Vienna, 786 p.

JEFFRIES, D. F.; PRESTON, A.; STEELE, A. K. 1973. Distribution of cesium-137 in British coastal waters. *Mar. Pollut. Bull.,* vol. 4, p. 118–22.

JOSEPH, A. B.; GUSTAFSON, P. F.; RUSSEL, I. R.; SCHUERT, E. A.; VOLCHOK, H. I.; TAMPLIN, A. 1971. Sources of radioactivity and their characteristics. In: *Radioactivity in the marine environment,* p. 6–41. Washington, D.C., National Academy of Sciences.

KOIDE, M.; GRIFFIN, J. J.; GOLDBERG, E. D. 1975. Records of plutonium fallout in marine and terrestrial samples. *J. Geophys. Res.,* vol. 80, p. 4153–62.

KREY, P. W. 1967. Atmospheric burnup of a plutonium-238 generator. *Science,* vol. 158, p. 769–71.

MARTELL, E. A. 1963. On the inventory of artificial tritium and its occurrence in atmospheric methane. *J. Geophys. Res.,* vol. 68, p. 3759–69.

NAS. 1975. *Assessing potential ocean pollutants.* Washington, D.C., National Academy of Science. 438 p.

NAS/NRC. 1957. *The effects of atomic radiation in oceanography and fisheries.* Washington, D.C., National Academy of Science/National Research Council. 137 p. (Publication No. 551.)

NOSHKIN, V. E. 1972. Ecological aspects of plutonium dissemination in aquatic environments. *Health physics,* vol. 22, p. 537–49.

NOSHKIN, V. E.; BOWEN, V. T.; WONG, K. M.; BURKE, J. C. 1972. Plutonium in North Atlantic Ocean organisms; ecological relationships. In: D. J. NELSON (ed.), *Radionuclides in Ecosystems, Proc. Third Nat. Symp. Radioecology, May 10–12, 1971, Oak Ridge, Tenn.,* vol. 2, p. 681–8.

NOVICK, S. 1974. Report card on nuclear power. *Environment,* vol. 16, no. 10, p. 6–12.

PRESTON, A. 1972. Artificial radioactivity in freshwater and estuarine systems. *Proc. Roy. Soc.* (London), vol. 180B, p. 421–36.

PRESTON, A. 1974. Artificial radioactivity in the sea. In: E. D. GOLDBERG (ed.), *The sea,* vol. 5, p. 817–36. New York, N.Y., Wiley-Interscience.

PRESTON, A.; FUKAI, R.; VOLCHOK, H. L.; YAMAGATA, N. 1971. Report of the Panel on Radioactivity. In: *Report of the Seminar on Methods of Detection, Measurement and Monitoring of Pollutants in the Marine Environment,* p. 87–99. Rome, FAO. (Marine Fisheries Reports, No. 99, Suppl. 1.)

ROETHER, W.; MÜNNICH, K. O. 1967. Cited in VOLCHOK *et al.,* 1971.

TEMPLETON, W. L.; NAKATANI, R. E.; HELD, E. E. 1971. Radiation effects. In: *Radioactivity in the marine environment,* p. 223–39. Washington, D.C., National Academy of Sciences.

VOLCHOK, H. L.; BOWEN, V. T.; FOLSOM, T. R.; BROECKER, W. S.; SCHUERT, E. A.; BIEN, G. S. 1971. Oceanic distributions of radionuclides from nuclear explosions. In: *Radioactivity in the marine environment,* p. 42–89. Washington, D.C., National Academy of Sciences.

WONG, K. M.; BURKE, J. C.; BOWEN, V. T. 1971. Plutonium concentrations in organisms of the Atlantic Ocean. *Health Physics Society Annual Symposium: Proceedings of Midyear Topical Symposium,* p. 529–39.

WONG, K. M.; HODGE, V. F.; FOLSOM, T. R. 1972. Plutonium and polonium inside giant brown algae. *Nature* vol. 237, p. 460–2.

WOODHEAD, D. S. 1973. Levels of radioactivity in the marine environment and the dose commitment to marine organisms. In: *Radioactive contamination of the marine environment,* p. 499–525. Vienna, International Atomic Energy Agency.

5.

Heavy metals

Introduction

Oceanic pollution by metals has been primarily observed in coastal waters as a consequence of river and industrial and domestic sewage discharges and of direct dumping of wastes. The one exception is lead, which appears to be primarily transported to the oceans via the atmosphere following its use as an antiknock agent in internal-combustion engine fuels (see Chapter 2).

Some metals are being mobilized about the airs and waters of the earth at rates comparable to, and sometimes exceeding, those in the weathering processes. This is especially evident for metals released to the atmosphere during the production of cement and the combustion of fossil fuels. For example, zinc released during fossil-fuel combustion totals 1 billion grammes per year, while somewhere between 3 and 30 billion grammes per year are mobilized by the major sedimentary cycle. The amounts of mercury released in the manufacture of cement appears to be nearly equivalent to that involved in the natural weathering cycle (see subsequent development in this chapter).

This chapter will deal with some of the general problems of metal pollution, with detailed analyses of three elements: mercury, cadmium and lead, which have been extensively studied in the past few years.

Speciation of heavy metals

Three chemical states of a heavy metal in sea water may be defined: particulate, colloidal and dissolved. In the dissolved state, the metal may be associated with ligands to form a variety of complexes. Although thermodynamic models have been devised to describe the distribution of a metal among various dissolved complexes, these have not been experimentally confirmed. The difficulties in the theoretical treatments are due to inadequate knowledge of the equilibrium constants between the metal and the different ligands, as well as to a lack of definitive values for the activities of the reacting species. Experimentally, the types of species in sea water are described in terms of the techniques employed—for example, dializable and non-dializable zinc; copper that passes through a 0.45 micron membrane filter and copper that does not.

An organism or a growing authigenic mineral may accumulate one form of a metal more or less readily than another form. Diatoms can accumulate iron in the particulate form but not as the complex of ethylene diamine tetra-acetic acid (EDTA) (Goldberg, 1952). The specific activity of radio-active zinc (radio-active decays per unit time per weight of stable zinc) in organisms differs from that in the water of their environment (Robertson *et al.*, 1968). The radio-active zinc is introduced in nuclear reactor effluent, and this suggests that the radio-active zinc is in different chemical forms from those of non-radio-active, naturally occurring zinc.

Experiments prompted by such observations have attempted to determine how rapidly the various species of a metal equilibrate under laboratory conditions. Piro *et al.* (1973) investigated the exchange reactions between stable ionic zinc and complexed zinc in sea water through polarographic assays. Zinc species were defined in the following way:

Ionic zinc, which corresponds to the amount detectable at a pH of 8 as the polarographic reduction wave.

Particulate zinc, which corresponds to the difference between the amounts detectable at a pH of 6 and at a pH of 8. The

particulate zinc is presumably solubilized by the acid treatment and converted to the ionic form.

Complex zinc, which corresponds to the difference between the amounts detectable between the values measured at pH 2 and pH 6. The acid treatment is presumed to convert the complexed zinc and the particulate zinc to ionic forms, amenable to polarographic measurement.

In natural sea waters, the zinc so defined is distributed between these three categories in the following way: 10–20 per cent in ionic form, including labile complexes; 30–50 per cent in particulate phases; and 40–50 per cent in complexed forms. The second and third forms can be transformed into ionic zinc by lowering the pH in times of about one minute. On the other hand, returning to a pH of 6 does not result in a rapid reformation of the complex species; reformation takes approximately ten hours. The re-establishment of equilibria between the three species at pH 8 takes even longer, about twenty-four hours.

Further evidence of the marked influence of pH upon the rates of transformation is shown in the work of Piro *et al.* (1973). If the ionic fraction is eliminated at a pH of 6, the complex and ionic species re-equilibrate in a few hours. If after the elimination of the ionic fraction at pH 6, the solution is slowly taken to a pH of 8, the re-establishment of equilibria between all three forms takes twenty-four hours. However, if at pH 8, the ionic form is removed, even after two days, an equilibrium distribution among the three forms cannot be observed. The investigators suggest that within that period of time no zinc is released from either the particulate or the complexed forms.

Radio-active zinc (presumably entirely in the ionic form) when added to sea water at a pH of 8 did not equilibrate or even transform to the complex forms but partitioned itself between the ionic and particulate phases after one year. Marine organisms preferentially accumulate zinc in the ionic form, according to Bernhard and Zattera (1969). Since radio-active wastes contain primarily ionic zinc, the organisms can be expected to have a higher specific activity than that of the total zinc in their surrounding waters. Zinc added in ionic form to sea waters will probably distribute itself primarily between the ionic and particulate phases, not immediately entering into association with complexing agents. Other heavy metals probably have unique but comparable behaviours, a knowledge of which is important in understanding the fates of wastes containing these metals in the ocean system.

Micro-organisms can convert toxic species of metals to non-toxic forms as well as transforming non-toxic forms into toxic ones (Wood, 1974). Mercury provides a simple example, where the disproportionation reaction $Hg_2^{++} = Hg^{++} + Hg^0$ can be affected by micro-organisms. Aerobes can solubilize HgS by oxidizing the sulphide to sulphate with the consequent production of Hg^{++}. The divalent mercury can be reduced to the elemental form by an enzyme which is present in a number of bacteria.

Wood argues that this conversion can be considered a detoxification inasmuch as the elemental form of mercury has sufficient vapour pressure to be lost from aqueous environments into the atmosphere. Some bacteria are able to convert the divalent mercury to methylmercury and dimethyl mercury, which are neurotoxins, through the use of methyl-B-12 compounds. Still other micro-organisms can convert methyl mercury to elemental mercury and methane, a detoxification reaction. Other elements like mercury —tin, palladium, platinum, gold and thallium—can be methylated in the environment (Wood, 1974). Some elements like lead, cadmium and zinc will not be methylated. The former group of elements,

if their chemical species in the oceans are methylated by appropriate micro-organisms, could be toxic to some living organisms.

The chemical forms of the heavy metals introduced to the oceans by man's activities can be quite different from those mobilized by natural processes. Thus, the fluxes and the residence times of the anthropogenic species of a given metal can differ from those of the natural species. For example, it appears that the mercury entering Minamata Bay in the discharged spent catalysts was in the form of methyl mercury, which is rapidly taken up by the resident organisms. The naturally occurring inorganic forms, which are less reactive, were not involved in the consequent tragedy. Radio-active zinc, formed in nuclear reactors by interactions of the neutrons with the components of the system does not rapidly equilibrate with the stable zinc species in sea water.

The chemical forms of anthropogenic elements can be better understood from investigations of the waters, sediments and organisms near the sites of discharge. We will examine two such investigations: one involving a sewer outfall, the other, a river. Bruland *et al.* (1974a) examined the association of heavy metals with various chemical phases in a deposit near a sewer outfall off southern California (Whites Point) and in adjacent sediments where influences of society were minimal (Soledad basin off Baja California). The components with which the heavy metals were associated were divided into three groups on the basis of chemical behaviours:

Reducible phases, extractable by 25 per cent acetic acid into a solution of hydroxylamine hydrochloride. Acid-soluble phases, such as carbonates or some sulphides, sorbed materials displaceable by hydrogen ions and some organic materials, will be affected by this treatment.

Oxidizable phases, extractable with 30 per cent hydrogen peroxide. Such materials include oxidizable organic matter.

The resistant residue surviving the above two treatments.

The distribution of heavy metals between these three components is given in Table 24 for sediments from the Soledad basin and from the Whites Point outfall off Los Angeles.

Most of the metals, lead, copper, chromium, zinc, silver, nickel and cobalt, are more readily solubilized by oxidative or reductive treatments in the outfall deposits than in the sediments presumably not influenced by man's activities. Manganese, a metal whose concentrations in the California basin deposits are not affected measurably by man's activities, has similar distributions among the components in both sediments. So does vanadium which is a more difficult case to interpret, since it is an element with a substantial anthropogenic flux. A simple explanation to account for the high concentrations of vanadium in the resistant phases of both

TABLE 24. Distribution of heavy metals between sedimentary phases of Whites Point Sewer Outfall deposits and of the Soledad basin

Element	Phase	Soledad		Whites Point	
		%	p.p.m.	%	p.p.m.
Pb	Reducible	78	5.5	83	357
	Oxidizable	<5	<0.1	4	17
	Resistant	22	1.5	13	56
Cr	Reducible	19	22	49	358
	Oxidizable	42	49	37	270
	Resistant	40	47	14	102
Cu	Reducible	11	4	31	190
	Oxidizable	17	6	49	300
	Resistant	72	24	20	120
Zn	Reducible	18	13	85	1,530
	Oxidizable	18	13	8	144
	Resistant	64	46	7	126
V	Reducible	32	36	20	34
	Oxidizable	28	32	10	17
	Resistant	40	46	70	118
Ag	Reducible	49	1.0	70	14
	Oxidizable	<15	<0.3	9	2
	Resistant	36	0.7	20	4
Ni	Reducible	27	15	39	33
	Oxidizable	48	27	26	22
	Resistant	25	14	25	29
Co	Reducible	<4	<0.2	17	2.4
	Oxidizable	<10	<0.4	20	2.8
	Resistant	86	3.4	63	8.8
Mn	Reducible	7	12	11	42
	Oxidizable	11	19	15	58
	Resistant	82	139	74	286

Reproduced from K. W. Bruland, K. K. Bertine, M. Koide and E. D. Goldberg, 'History of Metal Pollution in Southern California Coastal Zone', *Environ. Sci. Technol.*, Vol. 8, 1974, p. 425–32.

deposits would relate to the resistant nature of the vanadium species introduced by man.

A study in the Rhine estuary by DeGroot (1973) proposed a model by which heavy metals entering that estuary and substantially deriving from industrial wastes, may end up as organic complexes in coastal waters. Rhine waters carry smaller burdens of some heavy metals than do Rhine sediments. For example, there is 2.9 times more mercury, 2.1 times more arsenic and 1.6 times more copper affixed to the sediments with a particle size less than 16 microns than in the water. However, the metals associated with the sediments are released to the sea water upon entering the estuarine zone. DeGroot proposes that the cause of this mobilization is the formation of complexes between the metal ions sorbed to the sediment and organic ligands released in the intensive decomposition of organic matter in the sediments. The degree of mobilization varies according to the metal, with over 80 per cent of the mercury, copper, zinc and lead leaving the solid phases and entering the liquid phase. It tends to follow the Irving-Williams order where the relative stabilities of metal complexes take the order: $Fe(III) > Hg(II) > Cu(II) > Pb(II) > Zn(II) > Co(II) > Mn(II)$. The trivalent species, samarium, scandium and lathanum as well as manganese, are only slightly removed from the solid phases. DeGroot indicates that the organic phases responsible are the fulvic acids. The acids with molecular weights between 1,000 and 10,000 were more effective in complexing than those with molecular weights less than 1,000 or greater than 10,000. The functional groups involved were carboxyls, phenolic hydroxyls and carbonyls.

The subsequent fate of these organic complexes is not known. Also, the role of inorganic complexing agents, such as chloride or sulphate, has not been compared quantitatively to that of the fulvic acids and other organic compounds. Perhaps the mercury-chloride complexes are significant in the removal of mercury from sorption sites on the sediments. In any case, DeGroot's model from the Rhine study suggests that anthropogenic metals entering coastal waters may form organic complexes with degradation products from biological phases and may have different environmental behaviours from those of their naturally occurring counterparts in the ocean. Other estuarine areas would need to be investigated to substantiate the extent to which these processes occur elsewhere.

Anthropogenic fluxes

The mobilization of certain heavy metals to the oceans via the atmosphere in amounts comparable to those moving to the oceans via rivers in the major sedimentary cycle is induced by two high-temperature industrial processes: the combustion of fossil fuels and the production of cement. Since the greater part of these industrial activities takes place in the mid-latitudes of the northern hemisphere changes in heavy metal compositions of natural waters and of air will be most evident at these latitudes. As previously pointed out, the higher sulphate levels in rivers of the northern hemisphere, compared to those of the southern in societies not yet technologically advanced, are attributed to fossil-fuel burning.

Estimates of the fossil-fuel burning fluxes of heavy metals have been made by Bertine and Goldberg (1971). Their model utilized the consumption of fossil fuels in 1967: coal, 1.75×10^{15} g; lignite, 1.04×10^{15} g; fuel oils, 1.63×10^{15} g; and natural gas 0.66×10^{15} g. The literature was surveyed for reasonable values of elemental contents of these fuels, and estimates of the amounts released to the atmosphere

were based upon the following assumptions: the fly ash released to the atmosphere from the burning of coals and oils is about 10 per cent of the total ash; and 50 per cent of the coal is used in the manufacture of coke. These results are given in Table 25. For such elements as barium and mercury, fossil-fuel mobilization to the atmosphere appears to be within an order of magnitude of the river fluxes of these metals to the oceans. The materials mobilized in these two ways enter the oceans at different places—the rivers bring their metal loads to the coastal ocean where they can be rapidly involved in biological and inorganic chemistries; the winds move the fly ash and gaseous metal phases to the open as well as the coastal ocean.

Selective volatilization can introduce the readily distillable materials into the atmosphere in higher concentrations than those indicated in Table 25. Bertine and Goldberg (1971) suggest that those elements that will be mobilized more effectively through volatilization processes may be found in emissions from the d.c. arc as observed by spectrographers. On such a basis they suggest a preferential transfer, above that indicated in the table, for arsenic, mercury, cadmium, tin, antimony, lead, zinc, thallium, silver and bismuth.

Elemental carbon produced during fossil-fuel burning is detectable in the surface sediments of the Sierra Leone Rise in the Atlantic Ocean (Parkin *et al.*, 1970). Fly ash, consisting mainly of colourless glass spheres with occasional red, yellow or brown particles has been isolated from dusts of the prevailing westerlies and the northeast trade winds in the Atlantic. Thus, the combustion of coal, oil, and natural gas is producing materials evident in the winds and the sediments of the Atlantic.

Smaller, but similar types of emissions to the atmosphere are a consequence of cement production. In 1970 the world cement production was about 5.7×10^{14} grammes per year (USDI, 1972). About 95 per cent of the world production is Portland cement whose chemical formulation can be considered to be one-third shale and two-thirds limestone. The cement

TABLE 25. Amounts of elements mobilized into the atmosphere as a result of weathering processes and the combustion of fossil fuels (p.p.m.)

Element	Fossil-fuel concentration (p.p.m.)		Fossil-fuel mobilization ($\times 10^9$ g/year)			Weathering mobilization ($\times 10^9$ g/year)*	
	Coal	Oil	Coal	Oil	Total	River flow	Sediments
Li	65		9			110	12
Be	3	0.0004	0.41	0.00006	0.41		5.6
B	75	0.002	10.5	0.0003	10.5	360	
Na	2,000	2	280	0.33	280	230,000	57,000
Mg	2,000	0.1	280	0.02	280	148,000	42,000
Al	10,000	0.5	1,400	0.08	1,400	14,000	140,000
P	500		70			720	
S	20,000	3,400	2,800	550	3,400	140,000	
Cl	1,000		140			280,000	
K	1,000		140			83,000	48,000
Ca	10,000	5	1,400	0.82	1,400	540,000	70,000
Sc	5	0.001	0.7	0.0002	0.7	0.14	10

TABLE 25 (*continued*)

Element	Fossil-fuel concentration (p.p.m.)		Fossil-fuel mobilization ($\times 10^9$ g/year)			Weathering mobilization ($\times 10^9$ g/year)*	
	Coal	Oil	Coal	Oil	Total	River flow	Sediments
Ti	500	0.1	70	0.02	70	108	9,000
V	25	50	3.5	8.2	12	32	280
Cr	10	0.3	1.4	0.05	1.5	36	200
Mn	50	0.1	7	0.02	7	250	2,000
Fe	10,000	2.5	1,400	0.41	1,400	24,000	100,000
Co	5	0.2	0.7	0.03	0.7	7.2	8
Ni	15	10	2.1	1.6	3.7	11	160
Cu	15	0.14	2.1	0.023	2.1	250	80
Zn	50	0.25	7	0.04	7	720	80
Ga	7	0.01	1	0.002	1	3	30
Ge	5	0.001	0.7	0.0002	0.7		12
As	5	0.01	0.7	0.002	0.7	72	
Se	3	0.17	0.42	0.03	0.45	7.2	
Rb	100		14			36	600
Sr	500	0.1	70	0.02	70	1,800	600
Y	10	0.001	1.4	0.0002	1.4	25	60
Mo	5	10	0.7	1.6	2.3	36	28
Ag	0.5	0.0001	0.07	0.00002	0.07	11	0.03
Cd		0.01		0.002			0.5
Sn	2	0.01	0.28	0.002	0.28		11
Ba	500	0.1	70	0.02	70	360	500
La	10	0.005	1.4	0.0008	1.4	7.2	40
Ce	11.5	0.01	1.6	0.002	1.6	2.2	90
Pr	2.2		0.31			1.1	11
Nd	4.7		0.65			7.2	50
Sm	1.6		0.22			1.1	13
Eu	0.7		0.1			0.25	2.1
Gd	1.6		0.22			1.4	13
Tb	0.3		0.042			0.29	
Ho	0.3		0.042			0.36	2.3
Er	0.6	0.001	0.085	0.0002	0.085	1.8	5.0
Tm	0.1		0.014			0.32	0.4
Yb	0.5		0.07			1.8	5.3
Lu	0.07		0.01			0.29	1.5
Re	0.05		0.007				0.001
Hg	0.012	10	0.0017	1.6	1.6	2.5	1.0
Pb	25	0.3	3.5	0.05	3.6	110	21
Bi	5.5		0.75				0.6
U	1.0	0.001	0.14	0.001	0.14	11	8

* Two different techniques were employed to calculate the weathering fluxes of the metals—one based upon sedimentation and one upon river flow. The differences of the results between the two techniques usually reflect the inadequacies of the existing data.

Reproduced from K. K. Bertine and E. D. Goldberg, 'Fossil Fuel Combustion and the Major Sedimentary Cycle', *Science*, Vol. 173, 1971, p. 233–5.

is produced by roasting such a mixture at temperatures between 1,450° and 1,600° C in the burning zones of the kilns.

Particulate emissions from cement manufacturing have been estimated to range between 116 and 171 kg per ton in the United States (EPA, 1972). Applying a value of 150 kg per ton to world production, there are 860,000 tons of particulates emitted annually to the atmosphere from this activity.

An estimate of the volatile substances emitted from cement manufacture can be obtained in the following way. About 36 per cent of the ignition loss is in the liberation of carbon dioxide from limestone. Hence, the initial amount of limestone needed to produce the cement is 5.7 $\times 10^{14} \times 2/3 \times 100$ $CaCO_3/56$ g. $CaO = 6.8 \times 10^{14}$ grammes per year. The initial amount of shale is $1/3 \times 5.7 \times 10^{14}$ grammes per year $= 1.9 \times 10^{14}$ grammes per year.

The time of passage through a kiln is 2–4 hours. Those metals whose oxides have boiling points below 2,000° C may be expected to enter the gas phase in substantial amounts. Those below 1,500° C may be expected to be totally volatilized. As an approximation, we assumed those metal oxides with the following boiling points would be volatilized to the associated percentage of the amount in the cement components: <1,500, 100 per cent; 1,500–1,600, 50 per cent; 1,600–1,700, 40 per cent; 1,700–1,800, 30 per cent; 1,800–1,900, 20 per cent; 1,900–2,000, 10 per cent; >2,000, 0 per cent.

The data for a group of elements is given in Table 26. The mobilization to the atmosphere by cement production appears to be greater than the mobilization by the burning of fossil fuels for such elements as arsenic, boron, lead, selenium and zinc. This may be true for other metals for which data on boiling points are at present unavailable. In some cases, where the oxides decomposed upon heating below 2,000° C, the volatility of the metal was used. A substantial portion of the volatilized material will wash out into surface waters of the oceans. However, it is not clear what measurable effects, if any, they might have.

Specific uses of some metals can allow their entry into the marine environment in amounts near those introduced naturally during weathering cycles. For example, copper and cupric oxide are used in antifouling paints for the protection of marine

TABLE 26. Emissions of volatile oxides from the production of cement

Element	B.P. of oxide	Shales (p.p.m.)	Limestones (p.p.m.)	Grammes in 5.7 × 10^{14} grammes cement	Emission (grammes per year)
Sb	1,550	1.5	0.2	4.2×10^8	2.1×10^8
As	315	13	1	3.2×10^9	3.2×10^9
B	ca 1,860	100	20	3.3×10^{10}	3.3×10^{10}
Cd	d. 900–1,000	0.3	0.035	8.1×10^7	8.1×10^7
Cs	690 (metal)	6	6	5.2×10^9	5.2×10^9
Pb	1,744 (metal)	20	9	9.9×10^{10}	3.0×10^{10}
Li	1,200[600 mm Hg]	66	5	1.4×10^{10}	1.4×10^{10}
Hg	356 (metal)	0.4	0.04	1.0×10^8	1.0×10^8
Rb	d. 400	140	3	2.9×10^{10}	2.9×10^{10}
Se	subl. 350	0.6	0.88	7.1×10^8	7.1×10^8
Tl	1,457 (metal)	1.4	0	2.7×10^8	2.7×10^8
Zn	907 (metal)	95	20	3.2×10^{10}	3.2×10^{10}

ships. To prevent the attachment and subsequent growth of organisms, the copper must leach into the water coating the hull of the ship at a rapid enough rate to interfere with the organism's activity and yet not so rapid as to exhaust the copper in a short time. An effective rate is around 10 μg/cm^2 per day. Most anti-fouling paints contain about 100 to 200 grammes of cupric oxide per litre of paint.

An estimate of the amount of copper leaking into the oceans from its use as an anti-fouling agent may be made in the following way. There are about 20,000 merchant ships in the world (Anon, 1973) and about 15,000 naval vessels (ISS, 1972). We assume an average ship length of 100 metres and an average draft of 10 metres. Also we suggest that the two sides of a ship provide an area of $2 \times 100 \times 10 = 2,000$ m^2. For 35,000 ships, this is an area of 70×10^6 m^2. At an annual leaching rate of 10 μg/cm^2, this corresponds to a flux of 2.5×10^9 grammes of copper per year. The flux of copper from rivers is estimated to be 250×10^9 grammes per year. Although this anthropogenic flux is but 1 per cent of the natural flux, the influence of the leached copper may be evident in harbours, coastal area or shipping lanes.

Heavy metal distributions in coastal waters

Reactive elements in sea waters, such as the heavy metals, are rapidly removed from coastal waters to the sediments in organic precipitates or solid biological phases as a result of the intense chemical and biological activities. In the previous section we have noted that some of these metals enter the oceans adsorbed to solids and are subsequently released to sea water. Residence times for lead and radium in the biologically active upwelling waters of the

Gulf of California, for example, are estimated to be of the order of months (Bruland *et al.*, 1974*b*). Thus, in coastal regions adjacent to rivers or outfalls, onshore-offshore concentration gradients for the heavy metals would be expected. Such is the case, for example, in the North Sea (Preston, 1973*a*). In its southern areas, manganese, zinc, copper and nickel showed highest concentrations near shore, generally in the vicinity of estuaries. Cadmium did not display such patterns and Preston indicates that the contribution from terrestrial sources seems to be small over the area studied. Substantial fractions of each metal were associated with the particulate phases. The North Sea waters did not show as high upper limits of the heavy metal concentrations as those from the Irish Sea.

Similarly, Preston (1973*b*) finds the highest metal concentrations in inshore waters of other British coastal regions. The concentrations of such metals as zinc, iron, manganese, copper, nickel, lead and cadmium were not significantly different in open-ocean waters of the North Sea and Irish Sea compared to those in the adjacent open Atlantic Ocean, again emphasizing the site of the removal processes to be the innermost inshore waters. Here is where the most intensive soiling of the marine system is taking place.

Analyses of the heavy metals in the seaweed *Fucus* collected in 1961 and in 1970 suggested that there was little, if any, change in the heavy-metal concentrations in these waters examined by Preston during this time interval. The geometric mean for the ratio of the elemental concentrations in 1970 to 1961 was: cadmium, 0.68; copper, 0.90; iron, 1.03; manganese, 0.86; nickel, 0.94; lead, 0.82 and zinc, 0.90. There is a possibility that the concentrations of cadmium in the waters may have decreased.

In depth profiles of the near-shore waters no significant difference was evident

TABLE 27.

	Cd	Pb	Zn	Cu	Fe	Mn	Ni	Co
Enrichment, coal ash over sediments	54	43	25	20	7	12	7	13
Enrichment, surface over deep sediments	6.9	4.1	2.9	1.9	1	1	1	1

in metal concentrations in filtered sea-water samples between midwater, surface, and bottom, indicating that mixing was effective in these shallow sea areas (Preston, 1973a; 1973b).

Two recent studies have pinpointed anthropogenic metals that are accumulating in coastal-zone sediments as well as identified those whose concentrations appear unaffected by man's activities. Erlenkeuser et al. (1974) carried out studies on recent sediment cores of the western Baltic Sea, while Bruland et al. (1974a) investigated the deposits of the basins off southern California, adjacent to the highly industrialized Los Angeles area. Both groups found enhanced concentrations of lead, cadmium, zinc and copper, and unchanged concentrations of nickel, cobalt, manganese and iron in surface layers compared to those in deeper strata accumulated earlier. In addition, Bruland et al. (1974a) found that chromium, silver, vanadium and molybdenum belonged to the former group and aluminium to the latter. It must be emphasized that these sedimentary records of pollution are biased towards those species with short residence times in coastal waters. Chemicals that are less reactive and remain in solution do not readily reveal a pollution history in the deposits. Such may be the case with nickel and cobalt. Also, for those elements whose natural fluxes are substantial, such as iron and manganese, man's influence on the sedimentary record may be difficult to detect.

Erlenkeuser et al. (1974) determined their time scales with radiocarbon dates.

Rates of sedimentation were around 1.4 mm per year. Within the upper 20 cm of sediment cadmium, lead, zinc and copper increased seven-, four-, three- and two-fold, respectively, while iron, manganese, nickel and cobalt remained unchanged. The authors attribute the source of the increased heavy metals in the upper parts of the sediments to the combustion of coal. Their argument may be seen from the data given in Table 27.

To account for the enhanced levels of the cadmium, lead, zinc and copper in the upper levels of these Baltic sediments, only 7 per cent by weight of coal ash must be added. This amount is so small that the iron, manganese, nickel and cobalt concentrations would not be significantly affected.

The increases in the cadmium, lead, zinc and copper are seen to start about 1820. This observation is in accord with that of Murozumi et al. (1969) who noted that the Industrial Revolution could be detected in the Greenland ice sheets beginning around 1850. The increase in European coal production is shown in Figure 20 and parallels the increases in the anthropogenic input of heavy metals to these sediments.

Technologically advanced societies inject heavy metals into coastal oceanic areas through other activities than fossil-fuel burning. No matter what the source, however, the composition of the inshore sediments can be altered substantially, as can be seen in the results of Bruland et al. (1974a) in their studies of four basins off the coasts of North America. The recent rates of metal accumulation were compared with

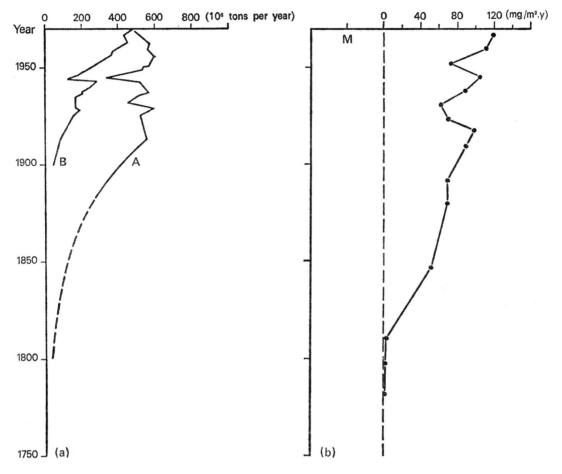

FIG. 20. (a) European coal production during the last 170 years: A, coal; B, lignite. (b) The anthropogenic input of metals to the Baltic Sea (Erlenkeuser *et al.*, 1974).

those of a hundred years ago (Table 28). The anthropogenic fluxes of lead, for example, are three times higher than the natural ones on the average, while those of chromium, zinc, copper, silver, vanadium, cadmium and molybdenum are of the same order of magnitude or somewhat less than the natural ones. For nickel, cobalt, manganese and iron there appear to be no measurable changes in their sediment concentrations attributable to the activities of society.

The sources and transport paths for these metals have not yet been established in detail. For lead, combustion as lead tetraethyl and lead tetramethyl in internal combustion engines is undisputed. For the other metals, multiplicity of uses is common. At the present time it is not possible

TABLE 28. Fluxes of heavy metals into sediments of California coastal basins

Element	Flux	Fluxes in $\mu g/cm^2/year$				
		San Pedro	Santa Monica	Santa Barbara	Soledad	Average
Pb	Anthropogenic	1.7	0.9	2.1		1.6
	Natural Rainfall	0.26	0.24	1.0	0.23	0.5
Cr	Anthropogenic	3.1	2.6	2.9		2.9
	Natural	2.8	2.1	10.7	4.6	5.2
Zn	Anthropogenic	1.9	2.1	2.2		2.1
	Natural Rainfall	3.1	2.8	9.7	2.8	5.2
Cu	Anthropogenic	1.4	1.1	1.4		1.3
	Natural Rainfall	1.2	1.0	2.6	1.4	1.6
Ag	Anthropogenic	0.09	0.09	0.10		0.09
	Natural	0.05	0.03	0.11	0.08	0.06
V	Anthropogenic	1.5	2.6	7.8		4.0
	Natural	3.5	3.4	13.6	4.6	6.8
Cd	Anthropogenic			0.07		0.07
	Natural			0.14		0.14
Mo	Anthropogenic		0.8			0.8
	Natural		0.08			0.08
Ni	Natural Rainfall	1.6	1.3	4.1	2.3	2.3
Co	Natural	0.33	0.26	1.0	0.17	0.53
Mn	Natural Rainfall	13	8	24	7	15
Fe	Natural	1,260	1,200	3,060	840	1,800
Al	Natural	1,740	1,630	4,860	1,280	2,700

Reproduced from K. W. Bruland, K. K. Bertine, M. Koide and E. D. Goldberg, 'History of Metal Pollution in Southern California Coastal Zone', *Environ. Sci. Technol.*, Vol. 8, 1974, p. 425–32.

to evaluate the contributions from different transporting agencies—the winds, sewer outfalls, storm runoff and river runoff. Independent estimates of the fluxes from these paths are usually of the same order of magnitude and the identification of the principal path is difficult (Bruland *et al.*, 1974*a*). Difficulties in assessment of existing data can be seen in Table 29 where flux estimates along five different paths have been made independently: the anthropogenic fluxes from the sediment record; waste-water fluxes from the composition of sewer outfall waters; storm runoff and dry-weather flow from direct measurements; and finally atmospheric washout fluxes from rainfall analyses. The large errors accompanying these values make comparisons meaningless at the present time, although in a few cases it appears that one or more transport paths can be regarded as insignificant.

TABLE 29. Fluxes of materials to southern California coastal region (in tons/year/12,000 km²)

Element	Flux of anthropogenic components to sediments	Waste waters	Storm water plus dry-weather flow	Washout fluxes
Pb	190	213	90	156
Cr	350	649	25	12
Zn	250	1,680	101	550
Cu	160	567	18	60
Ag	11	15	1	5
V	480	. . .	. . .	17
Cd	8	54	1	48
Mo	100	. . .	. . .	24

Reproduced from K. W. Bruland, K. K. Bertine, M. Koide and E. D. Goldsberg, 'History of Metal Pollution in Southern California Coastal Zone', *Environ. Sci. Technol.*, Vol. 8, 1974, p. 425–32.

Lead

The concentrations of lead, unlike those of any other heavy metal, have been altered in vast expanses of coastal waters as a result of the use of lead alkyls as antiknock additives in fuels of internal-combustion engines. Lead is introduced to the sedimentary cycle by man in amounts that rival those of natural processes. Murozumi *et al.* (1969) indicated that in 1966 the world lead production was 3.5 million tons per year with 3.1 million tons of this produced in the northern hemisphere. Somewhat less than 10 per cent of this total production, 0.31 million tons per year, was burned as lead alkyls, the bulk of which entered the atmosphere in exhausts from internal-combustion engines (see Chapter 2, page 42). The river influxes of soluble and particulate lead to the marine environment have been estimated as 0.24 and 0.50 million tons per year, respectively.

Up to the present no deleterious effects from the existing lead burdens in marine organisms have been demonstrated. There is a concern from the human-health standpoint of the consumption of lead-contaminated organisms even though they may be unaffected by high amounts of this metal. Analyses of lead in the Greenland ice sheet have indicated a five-hundredfold increase of lead fallout over the last 2,800 years (Murozumi *et al.*, 1969). The manner in which the smelting of lead and the production of lead alkyls has increased from 1750 to 1966 is presented in Table 30.

The residence time of lead in marine waters has been estimated at 400 years (Goldberg *et al.*, 1971). This relatively low value reflects its involvement in biological reactions and its uptake by marine organisms. The residence time in highly productive coastal waters has been estimated at a month (Bruland *et al.*, 1974a) further emphasizing the biological interactions of lead.

The entry of the lead aerosols into the coastal water of the Pacific, Atlantic and Mediterranean has significantly altered the surface sea-water concentrations of lead (Chow and Patterson, 1966). Whereas the concentration of barium, an element with a presumably similar environmental chemistry to that of lead, normally shows a steady increase with depth, the concentrations of lead are almost always higher in surface

TABLE 30. Amounts of lead smelted or burned as alkyl lead per year, 1750–1966 (in thousands of tons)

Year	Northern hemisphere			Southern hemisphere
	Primary smelting	Secondary smelting	Burned alkyl-lead	Primary smelting
1750	60			40
1800	90			50
1860	220			
1880	400			
1890	520			40
1900	750	0		80
1910	940	60		100
1920	880	200	0	110
1930	1,200	400	4	170
1940	1,300	400	36	230
1950	1,300	550	110	240
1960	1,900	600	180	360
1966	2,400	700	310	350

Reproduced from J. Murozumi, T. J. Chow and C. Patterson, 'Chemical Concentrations of Pollutant Lead Aerosols, Terrestrial Dusts and Sea Salts in Greenland and Antarctic Snow Strata', *Geochim. Cosmochim. Acta*, Vol. 33, 1969, p. 1247–94.

TABLE 31.

Source	Ratio of lead-206 to:		
	Lead-204	Lead-207	Lead-208
California gasolines-average of those sold in 1964	17.92	1.145	0.4728
Of those sold in 1968	18.08	1.155	0.4756
San Pedro sediment, 0–2 cm	18.36	1.176	0.4782
San Pedro sediment, 30–32 cm	19.11	1.208	0.4845
Whites Point, 0–1 cm	18.09	1.166	0.4801

waters. This behaviour is not consistently shown by any other heavy metal. The Pacific and Mediterranean coastal surface waters have been affected to depths of about 500 metres, while the impact upon Atlantic surface waters has been evident only in shallow zones (Fig. 21). Chow and Patterson indicate an average lead concentration today in northern hemispheric coastal waters of 0.07 µg/kg, compared to estimated values of 0.01–0.02 µg/kg for times prior to the widespread use of lead alkyls. However, the concentrations reported nowadays are perhaps too high due to contamination during sampling or processing (C. Patterson, personal communication).

As we have seen in Chapter 2, coastal sediments maintain a record of lead fluxes to the coastal environment through an accommodation of a part of the lead in their solid phases. One characteristic of the lead,

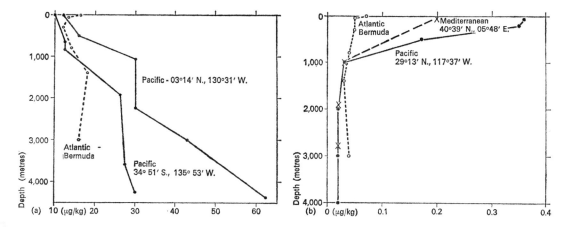

Fig. 21. (a) Barium in sea water; (b) lead in sea water (Chow and Patterson, 1966).

its isotopic composition, can be indicative of its origin (Chow *et al.*, 1973), as the data in Table 31 suggest.

The lead in petrol is usually derived from tertiary or older lead ores and possesses a quite distinct isotope composition. The Whites Point sediments, adjacent to a sewer outfall, appear to contain lead primarily derived from petrol. The San Pedro sediments have the isotopic compositions of their recently deposited lead influenced by petrol lead. Weathering lead is represented by that in the 30–32 cm stratum which accumulated before 1800.

Mercury

The environmental chemistry of mercury can readily be understood in terms of its unique physical and chemical properties. First of all, elemental mercury and many of its compounds have vapour pressures sufficiently high to cause significant movement of this metal in the vapour phase. Also, some of its compounds, such as the sulphides and oxides, are extremely insoluble. Mercury can be methylated by micro-organisms. These organic forms are the principal ones in many organisms. In addition they are much more toxic to man than inorganic mercury compounds.

The principal movements of mercury about the surface of the earth are initiated by a crustal degassing process (Weiss *et al.*, 1971). The calculation of mercury fluxes from the lithosphere to the atmosphere is based upon the element's concentrations in glacial waters whose rates of accumulation are known, and upon its contents in rains and airs in accordance with the washout models discussed in Chapter 2. Weiss *et al.* (1971) indicate global degassing rates of the continents within the range of 2.5×10^{10} to 1.5×10^{11} grammes per year. These values may be compared with the estimated river flux of mercury to the ocean of less than 3.8×10^9 grammes per year.

The world production of mercury in 1968 was 8.8×10^9 grammes per year, greater than the river load to the oceans but less than the degassing flux. Pockets of local pollution exist in coastal areas of the world ocean as a consequence of the release of the metal, primarily from industrial activities. However, the measurable alteration of mercury levels in the open ocean by man does not appear likely. Consider the volume of the open ocean mixed layer as 5×10^{19} l and the mercury level therein as 25 ng/l. This gives a total mercury content of the world ocean mixed layer of 1.25×10^{12}, over two orders of magnitude greater than the annual production by man.

One would not expect an alteration in the mercury contents of pelagic organisms as a result of man's use of the metal. This seems to be borne out by recent studies. Analyses of museum specimens of tuna caught sixty-two to ninety-three years ago and a swordfish caught twenty-five years ago reveal mercury levels similar to those of their living counterparts (Miller *et al.*, 1972). Barber *et al.* (1972) studied the mercury concentrations in bottom-dwelling fish and found a correlation between size and concentration, with the larger individuals having mercury concentrations up to 0.8 p.p.m. (wet weight). The 90-year-old specimens fit closely the size-concentration regression curve of the nine recently caught individuals of the same species. Thus, this study concluded that there was no man-caused mercury contamination in these fish.

The mercury in the sea water probably exists as chloro-complexes such as $HgCl_4^{--}$ or $HgCl_2^0$ with a residence time calculated to be 80,000 years. Mercury enters the food web at the level of micro-organisms, where the conversion to methyl mercury takes place as a detoxification reaction (Jernelöv, 1974). Marine plankton has mercury levels of 2–10 p.p.b., most of which is in the inorganic form while fish have concentrations varying from 0.01 to 2 p.p.m., primarily in the form of methyl mercury (Jernelov, 1974).

Jernelöv (1974) indicates that the mercury in the fish is bound to protein and not

dissolved in the fat as are the halogenated hydrocarbons. As a consequence, it is much more related to the size or age of the fish, as well as its trophic level. The highest mercury contents are found in the top-predators of the marine food web, such as the tuna and swordfish. Further, in general, for a given species the larger or older the fish the more mercury it will contain per unit weight.

While fish have most of their mercury in the methylated form, some marine mammals such as the seals contain very high amounts of mercury in an inorganic form, especially in their livers. Jernelöv (1974) suggests that the seals probably obtain their mercury in the methylated form from the fish they consume as food but later possibly break the covalent bonds between the mercury and the carbon atoms through some enzymatic process.

Much of the mercury introduced into the coastal ocean is probably retained there. Biological activity is high and mercury probably becomes associated with the biosphere. As a consequence of the net conveyance of biological phases from the surface to deeper waters and to the sediments, mercury is transported to the coastal deposits, many of which become anoxic with time. In the presence of sulphide ions, mercury is precipitated as the highly insoluble mercury sulphide. The possibility exists that such deposits, if they encounter oxidizing conditions, i.e. contact with aerated waters, will release mercury following oxidation of the sulphide to sulphate.

Correlations between heavy-metal concentrations in marine organisms have been pointed out in the literature, but seldom is any causal relationship established between them. In the case of mercury, however, a strong covariance has been observed with selenium in the livers of marine mammals (dolphins, porpoises and seals) which may be related to the protective effect given by selenium against the toxic action of mercury compounds (Koeman *et al.*, 1973). The co-variance extended over 2.5 orders of magnitude with a regression curve given by (Hg) p.p.m.=2.91 (Se) p.p.m.+5.8 p.p.m.

Only a small part of the mercury is in the form of methylmercury in either the brain or liver (2–14 per cent of the total mercury). The mercury and selenium occur chiefly in the nuclei and cell wall material. The mercury is strongly bound, as shown by the experimental fact that 90 per cent of the mercury remains attached to the tissue components after a benzene extraction of acidified homogenates.

Koeman *et al.* (1973) accept a previous hypothesis that mercury and selenium occur together in animal tissues and are associated with proteins by means of sulphur. The antagonism of selenium to toxic mercury compounds is inferred from protective effects observed in rats and quail.

Cadmium

The epidemic of cadmium poisoning (*Itai-itai* disease) in Japan following consumption of rice polluted with wastes from a zinc smelter has focused attention upon the environmental behaviour of this element. Although the marine system has not yet been implicated in any cadmium poisoning, there is concern about the possible build-up of cadmium in marine organisms through environmental releases by man.

The world production of cadmium in 1970 is estimated to be 16 million tons (USDI, 1972). The primary use of cadmium is in electroplating and probably accounts for 45–60 per cent of the annual demand (USDI, 1972). Some compounds of cadmium are employed as colourants and stabilizing agents in the manufacture of vinyl plastics. It is also used in primary batteries, in phosphors for television tubes and night-seeing devices, as a hardener in copper-cadmium alloys, in fusible alloys and solders, and as electrical contacts for switches and relays.

Cadmium enters the marine environment as a result of man's activities both through the atmosphere and through the hydrosphere. Davis *et al.* (1970) estimate that the United States injects about 2,300 tons annually into the atmosphere. Ninety per cent of this injection is a consequence of the volatilization of the metal which is produced during the smelting of sulphide ores, used in the manufacture of metallic alloys, or reprocessed from cadmium-plated materials or cadmium-containing alloys. Since United States production of cadmium is 26.9 per cent of the world production, one can assume that 8,600 tons of cadmium enter the atmosphere from its use by all nations. Fleischer *et al.* (1974) consider this estimate too high and propose an air injection of 1,680 tons annually with the following sources: primary cadmium production, 930 tons; coal and oil combustion, 120 tons; losses during use and disposal of cadmium-plated metals, 500 tons; plastic and pigments, 90 tons; and alloys and batteries, 40 tons.

Fleischer *et al.* (1974) estimate that 4,390 tons of cadmium per year enter terrestrial waters, with 3,000 tons per year originating from mining and ore-concentration processes, 240 tons per year from primary cadmium production; and 300 tons per year as wastes from electroplating and plastics formulation. To complete the flow sheet, 4,390 tons per year are accumulated in service and 2,180 tons per year are accumulated in land disposal sites (slag pits, land fills and mine tailings).

The cadmium content of sea waters ranges between 0.05 and 0.2 μg/l with an average near 0.15 (Fleischer *et al.*, 1974). Thus, the upper layers (100 m) of the world oceans contain 4.2×10^{12} grammes of cadmium, about one-fourth of the world's annual production. The total cadmium loss to the atmosphere and hydrosphere from the activities of man is 5.2×10^9 grammes per year and this could readily be accommodated in the surface layer without detection by existing techniques. The problem of a global surface-water ocean pollution by cadmium does not seem probable for the near future.

The total amount of cadmium introduced to the oceans by rivers in the major sedimentary cycle has been estimated to be 0.5×10^9 grammes per year, substantially less than the amount mobilized by man to the atmosphere and to the terrestrial waters. The residence time is estimated to be 500,000 years. Thus, any initial problems arising from cadmium pollution would be expected in coastal waters. In British coastal water, as noted previously, the cadmium concentrations may have decreased between 1961 and 1970 (Preston *et al.*, 1972).

Overview

The evident pollution of marine waters by heavy metals is in the coastal zone, where entry is primarily through industrial and/or domestic sewers, rivers and dumping. Where mixing of these waters with those of the open ocean is slow, build-up of heavy metal concentrations over natural levels can take place. The Minamata Bay disaster resulting from the dumping of spent wastes containing mercury evolved from such a set of events.

So far, the activities of man have altered the concentrations of only one heavy metal in coastal ocean waters significantly. Lead levels in surface waters have increased over the past decades through dispersal of the element in internal-combustion engine exhausts and subsequent entry to the marine system through atmospheric fallout and coastal runoff. Other trace metals may have enhanced concentrations in surface waters of the world ocean, especially in northern hemispheric waters, due to the combustion of fossil fuels and the production of cement, and perhaps other industrial activities. However, such changes have not yet been detected, perhaps as a result of the inadequate analytical techniques available over the past years. No adverse effects from lead or other heavy metals have yet been reported in marine organisms.

Many of the reactive heavy metals are rapidly removed from coastal waters to the underlying sediments with biological and inorganic solid phases. As a consequence, strata of these deposits may yield records of the intensities of heavy metal pollution. But of greater importance is the lack of build-up in coastal waters where they might play a determinate role in biological processes.

The management of the metals with regard to marine pollution necessitates an assessment of the heavy metals fluxes entering the oceans from the atmosphere, from outfalls, from river runoff and from storm runoff. The induction that much of the heavy metal pollution of the Baltic Sea is associated with the combustion of fossil fuels provides a most important guide to the formulation of policy to regulate influx, if the metals are shown to jeopardize the resources of the Baltic.

The chemical form of man-mobilized heavy metals is important in understanding their environmental behaviours and in making predictive models of their distributions. Present evidence suggests that often the natural dissolved forms differ from those introduced as pollutants. Only a very small amount of pertinent research in this area has so far been undertaken.

Bibliography

ANON. 1973. Merchant shipping data. *U.S. Naval Institute Proceedings*, vol. 99, p. 364.

BARBER, R. T.; VIJAYAKUMAR, A.; GROSS, F. A. 1972. Mercury concentrations in recent and ninety-year-old benthopelagic fish. *Science*, vol. 178, p. 636–8.

BERNHARD, M.; ZATTERA, A. 1969. A comparison between the uptake of radioactive and stable zinc by a marine unicellular alga. In: D. J. NELSON and F. C. EVANS (eds.), *Symposium on Radioecology*, p. 389–98. Washington, D.C., Clearing House Fed. Sci. Techn. Information, United States Department of Commerce. (Conf. 670503 (TID-4500).)

BERTINE, K. K.; GOLDBERG, E. D. 1971. Fossil fuel combustion and the major sedimentary cycle. *Science*, vol. 173, p. 233–5.

BRULAND, K. W.; BERTINE, K. K.; KOIDE, M.; GOLDBERG, E. D. 1974*a*. History of metal pollution in southern California coastal zone. *Environ. Sci. Technol.*, vol. 8, p. 425–32.

BRULAND, K. W.; KOIDE, M.; GOLDBERG, E. D. 1974*b*. The comparative marine geochemistries of lead-210 and radium-226. *J. Geophys. Res.*, vol. 79, p. 3083–6.

CHOW, T. J.; PATTERSON, C. C. 1966. Concentration profiles of barium and lead in Atlantic waters off Bermuda. *Earth Planet. Sci. Letters*, vol. 1, p. 397–400.

CHOW, T. J.; BRULAND, K. W.; BERTINE, K. K.; GOLDBERG, E. D. 1973. Lead pollution: records in southern California coastal sediments. *Science*, vol. 181, p. 551–2.

DAVIS, W. E. *et al.* 1970. *National inventory of sources and emission, 1968. Cadmium, Report to National Air Pollution Control Administration, Washington, D.C.*, p. 1–44.

DEGROOT, A. J. 1973. Occurrence and behavior of heavy metals in river deltas with special reference to the Rhine and Ems Rivers. In: E. D. GOLDBERG (ed.), *North Sea science*, p. 308–25. Cambridge, Mass., Massachusetts Institute of Technology Press.

EPA. 1972. *Compilation of air pollutant emission factors*. Research Triangle Park, North Carolina, Office of Air Programs.

ERLENKEUSER, H.; SEUSS, E.; WILLKOMM, H. 1974. Industrialization affects heavy metal and carbon isotope concentrations in recent Baltic Sea sediments. *Geochim. Cosmochim. Acta*, vol. 38, p. 823–42.

FLEISCHER, M.; SAROFIM, A. F.; FASSETT, D. W.; HAMMOND, P.; SHACKLETTE, H. T.; NISBETT, I. C. T.; EPSTEIN, S. 1974. Environmental impact of cadmium: A review by the panel on hazardous trace substances. *Environmental Health Perspectives*, vol. 7, p. 253–323.

GOLDBERG, E. D. 1952. Iron assimilation by marine diatoms. *Bio. Bull.*, vol. 102, p. 243–8.

GOLDBERG, E. D.; BROECKER, W. S.; GROSS, M. G.; TUREKIAN, K. K. 1971. Marine chemistry. In: *Radioactivity in the marine environment*, p. 137–46. Washington, D.C., National Academy of Science.

ISS. 1972. *Janes weapons systems, 1972*. London, Institute for Strategic Studies. 586 p.

JERNELOV, A. 1974. Heavy metals, metalloids and synthetic organics. In: E. D. GOLDBERG (ed.), *The sea*, vol. 5, New York, N.Y., Wiley-Interscience.

KOEMAN, J. H.; PEETERS, W. H. M.; KOUDSTAAL-HOL, C. H. M. 1973. Mercury-selenium correlation in marine mammals. *Nature*, vol. 245, p. 385–6.

MILLER, G. E.; GRANT, P. M.; KISHORE, R.; STEINKRUGER, F. J.; ROWLAND, F. S.; GUINN, V. P. 1972. Mercury concentrations in museum specimens of tuna and swordfish. *Science*, vol. 175, p. 1121–2.

MUROZUMI, J.; CHOW, T. J.; PATTERSON, C. 1969. Chemical concentrations of pollutant lead aerosols, terrestrial dusts and sea salts in Greenland and Antarctic snow strata. *Geochim. Cosmochim. Acta*, vol. 33, p. 1247–94.

PARKIN, D. W.; PHILLIPS, R.; SULLIVAN, R. A. L. 1970. Airborne dust collections over the North Atlantic. *J. Geophys. Res.*, vol. 75, p. 1782.

PIRO, A.; BERNHARD, M.; BRANICA, M.; VERZI, M. 1973. Incomplete exchange reaction between radioactive ionic zinc and stable natural zinc in sea water. In: *Radioactive contamination of the marine environment*, p. 29–45. Vienna, International Atomic Energy Agency.

PRESTON, A. 1973*a*. Trace metals in the North Sea. *Mar. Pollut. Bull.*, vol. 4, p. 135–8.

PRESTON, A. 1973*b*. Heavy metals in British waters. *Nature*, vol. 242, p. 95–7.

PRESTON, A.; JEFFERIES, D. F.; DUTTON, J. W. R.; HARVEY, B. R.; STEELE, A. K. 1972. British Isles coastal waters: the concentrations of selected heavy metals in sea water, suspended matter and biological indicators. A pilot survey. *Environ. Pollut.*, vol. 3, p. 69–82.

ROBERTSON, D. E.; RANCITELLI, L. A.; PERKINS, R. W. 1968. Multielement analysis of sea-water, marine organisms, and sediments by neutron activation without chemical separation. *Proc. Intern. Symp. on the Application of Neutron Activation Analysis in Oceanography*, p. 143–212. Brussels, Inst. Royal Sci. Naturelles de Belgique.

USDI. 1972. *Minerals yearbook, 1970*, vol. 1. Washington, D.C., United States Department of the Interior. 1,235 p.

WEISS, H. V.; KOIDE, M.; GOLDBERG, E. D. 1971. Mercury in a Greenland ice sheet: evidence of recent input by man. *Science*, vol. 174, p. 692–4.

WOOD, J. M. 1974. Biological cycles for toxic elements in the environment. *Science*, vol. 183, p. 1049–52.

6.   Petroleum hydrocarbons

Introduction

The spreading of petroleum hydrocarbons about the ocean system as a result of man's need for energy is evident through many visible displeasures: the soiling of beaches, the staining of surface waters with films and tarballs and the occurrences of dead or moribund birds. The chemical analyst is also aware of this dispersion. Marine organisms now carry body burdens of petroleum hydrocarbons disseminated by man. Surface waters of the Baltic have petroleum concentrations ranging between 0.3 and 1.0 mg/l (Simonov and Justchak, 1972).

An assessment of the seriousness of the hydrocarbon pollution problem is made difficult in a number of ways. First of all there are the analytical problems. Crude oils are composed of many tens of thousands of compounds and vary in composition ac-cording to their source. There are no generally accepted techniques for their assay. As a result, there are very few results for the petroleum content of ocean waters, organisms or sediments.

There are three general sources of petroleum-containing hydrocarbons to the ocean system: man-generated substances; hydrocarbons produced by the marine organisms; and those hydrocarbons that seep in naturally through the sea floor. The origins of any one sample of petroleum hydrocarbon from the marine environment can be difficult to determine.

Because of their complex nature and of the difficulties in analysis, studies upon the effects of low levels of petroleum hydrocarbons upon life processes have been limited. Although a sense of the role of these compounds in altering behaviour patterns through interference in chemoreception, in reducing the reproductive success of animals, or in the induction of carcinomas has been reached, results are still fragmentary and present conclusions are only tentative.

Future marine pollution by petroleum hydrocarbons appears to be less of a problem than that caused by the other materials of concern today. The petroleum reserves are estimated to be able to handle man's energy and chemical needs for at best a century or so. However, since the ocean pollution created by man's promiscuous release of petroleum hydrocarbons may prevail for a longer period of time, there is an immediate need to recognize low-level long-term effects upon life processes.

CRUDE OILS

The crude oils from which petroleum products are derived consist of 50–98 per cent hydrocarbons and the remainder consists of compounds containing oxygen, nitrogen and sulphur. The hydrocarbons can be grouped into three classes:

Alkanes extend from methane (CH_4) and ethane to compounds with sixty carbon atoms or more, such as n-hexacontane ($C_{60}H_{122}$, a microcrystalline wax). They can be straight chained or branched with the former class being more abundant.

Cycloalkanes (naphthenes) are five or six carbon atom rings such as the monocyclic cyclopentane or cyclohexane as well as some polycyclic substances. Alkyl substitutions on the rings are common.

Aromatic compounds are present in small amounts and include benzene and alkylbenzenes such as toluene and xylenes. In addition there are the polynuclear aromatics in the higher boiling fractions and they encompass two general types: the fused-ring compounds such as the alkyl naphthalenes and the linked rings such as the biphenyls. The former are usually more abundant. Also included in this class are the naphtheno-aromatics, molecules part aromatic and part cycloalkane.

One characteristic of crude oils that distinguishes them from biogenous hydrocarbons in the marine system is that they do not contain alkenes (olefinic hydrocarbons).

The total oxygen content of a crude oil may attain 2 per cent. The principal species include phenols (cresols and higher boiling point alkylphenols) and carboxylic acids, both straight chain and linked, such as hexanoic acid and 3-methyl pentanoic acid. Nitrogen varies between 0.05 and 0.8 per cent in crude oils primarily in pyridine and quinolines. Sulphur exists both as the element and in compounds from trace amounts to 5 per cent by weight in crudes. These compounds include hydrogen sulphide, mercaptans and aliphatic and cyclic sulphides.

The metallo-organics encompass the nickel and vanadium porphyrin complexes which allow these metals to reach concentrations of 5–40 p.p.m. In addition, there are very small concentrations of other metals such as iron, sodium and zinc, whose associations with other elements are still undetermined.

REFINED PETROLEUM PRODUCTS

The distillation of crude oils produces a variety of refined products which have been categorized in the following way:

Straight-run petrol, which boils at temperatures up to 200° C and contains compounds with 4 to 12 carbon atoms.

Middle distillate, with compounds of 12 to 20 carbon atoms and a boiling range of 185°–345° C. Such products as kerosene, heating oils, diesel oils and jet, rocket and gas turbine fuels come from this group.

Wide-cut gas oil with 29 to 36 carbon atoms and boiling points between 345° and 540° C. Waxes, lubricating oils and starting materials for the production of gasoline by catalytic cracking processes come from this group.

Residual oils, usually asphaltic in nature.

The alkenes may be found in concentrations of up to 30 per cent in petrol and about 1 per cent in jet fuel. The exact composition of any given refined product depends upon the character of the crude oil from which it was derived and upon the nature of the distillation process.

BIOGENOUS HYDROCARBONS

Hydrocarbons are found in terrestrial and marine organisms. They can be synthesized by organisms, obtained through food, or altered following ingestion. The following review is derived from NAS (1975).

Normal alkanes, primarily with odd-numbered carbon chains, are synthesized by both marine and by terrestrial organisms. Often one or two compounds, usually with an odd number of carbon atoms, predominate. In marine phytoplankton, alkanes with 15, 17, 19 and 21 carbon atoms are most abundant, while in marsh grasses and *Sargassum* C_{21} to C_{29} normal alkanes have the highest concentrations. Some bacteria have equal amounts of even and odd numbered normal alkanes with 25 to 32 carbon atoms.

Branched alkanes also are found in organisms. The most notable is pristane, found in fishes and apparently entering in the food. Besides pristane, several monomethyl branched alkanes have been identified.

There are representatives of the alkenes (olefines) in the biosphere, but as mentioned previously, not in crude oils. Squalene is the primary hydrocarbon in shark and cod livers. Isoprenoid C_{19} and C_{20} mono-, di-, and tri-olefines occur in copepods and some fish. In addition, straight chain mono- to hexa-olefines have been found in organisms of the sea. The NAS (1975) report cites no definite reports of occurrences of cycloalkanes, or of aromatic hydrocarbons being synthesized by marine organisms. The cycloalkene carotene is synthesized by marine organisms.

For most investigations of hydrocarbon pollution, it is necessary to be able to distinguish between natural hydrocarbons and those released to the environment through man's use of petroleum In addition, it would be valuable to identify substances synthesized by land plants which in one way or the other reach the oceans.

NAS (1975) indicates some guidelines to the differentiation of petroleum and biogenous hydrocarbons, noting that the following differences may not apply to all organisms, crude oils or refined products:

Petroleum contains a much more complex mixture of hydrocarbons with greater ranges of molecular structure and molecular weight.

Petroleums often contain homologous series of hydrocarbons where adjacent members with even and odd numbers of carbon atoms are present in nearly the same concentrations. This is not true for biosynthesized hydrocarbons.

Petroleum contains more kinds of cycloalkanes and aromatic hydrocarbons. Alkyl substituted ring compounds have not been reported in marine organisms. Examples are the mono-, di-, tri- and tetra-methyl benzenes and the mono-, di-, and tetra-methyl naphthalenes.

Petroleum contains naphtheno-aromatic hydrocarbons not yet reported in marine organisms. Petroleums also contain heterocompounds containing asphaltic compounds and the nickel and vanadium porphyrins.

PRODUCTION

Crude oil is mined in countries which in general do not refine it or use it. As a consequence, there is a flow of the crude oil in ships or in pipelines from one country to another. The world production in 1971 of 2,478.4 million tons is shown in Table 32 by country and area. About one-third of the crude oil comes from the Middle East

TABLE 32. World oil production, 1971

Country/area	Tons (millions)	Country/area	Tons (millions)
North America		*Middle East*	
United States		Iran	226.2
Crude oil	473.2	Iraq	83.4
Natural-gas liquids	60.1	Kuwait	147.1
Canada	77.1	Neutral Zone	28.3
Mexico	24.9	Qatar	20.5
TOTAL	635.3	Saudi Arabia	223.4
		Abu Dhabi	44.9
Caribbean		Oman	14.4
Venezuela	187.3	Turkey	3.5
Colombia	11.6	Other Middle East	15.5
Trinidad	6.8	TOTAL	807.2
TOTAL	205.7		
		Africa	
South America		Algeria	35.9
Argentina	21.9	Libya	132.9
Brazil	8.3	Other North Africa	25.7
Other South America	6.6	Nigeria	74.1
TOTAL	36.8	Other West Africa	11.9
		TOTAL	280.5
TOTAL Western Hemisphere	877.8	*South-East Asia*	
		Indonesia	43.9
Western Europe		Other South-East Asia	11.9
France	1.9	TOTAL	55.8
Federal Republic of Germany	7.4	*U.S.S.R.*	372.0
Austria	2.5	*Eastern Europe and China*	42.0
Other Western Europe	7.1	*Other Eastern Hemisphere*	24.2
TOTAL	18.9	TOTAL	1,600.6
		World (excluding U.S.S.R., Eastern Europe and China)	2,064.4
		World	2,478.4

Reproduced from National Academy of Science, *Petroleum in the Marine Environment*, Washington, D.C., 1975, 107 p.

and slightly above one-quarter from North America. Table 33 shows the world oil consumption for 1971. Where the United States mines 533.3 million tons, it utilizes 715 million tons out of a world consumption of 2,396 million tons. Table 34 shows the imports and exports in 1971. Japan and Western Europe are the primary importers. The Middle East is the primary exporter.

TABLE 33. World oil consumption, 1971

Country/area	Tons (millions)	Country/area	Tons (millions)
United States	715	Middle East	54
Canada	77	Africa	45
Mexico	26	South Asia	29
Caribbean	56	South-East Asia	62
South America	66	Japan	220
TOTAL Western Hemisphere	940	Australasia	30
		U.S.S.R., Eastern Europe and China	364
Belgium and Luxembourg	29	TOTAL Eastern Hemisphere	1,456
Netherlands	36		
France	103	World (excluding U.S.S.R., Eastern Europe and China)	2,032
Federal Republic of Germany	133		
Italy	92	World	2,396
United Kingdom	103		
Scandinavia	54		
Spain	27		
Other Western Europe	75	Reproduced from National Academy of Science, *Petroleum in the Marine Environment*, Washington, D.C., 1975, 107 p.	
TOTAL Western Europe	652		

TABLE 34. Inter-area total oil movements 1971 (in millions of tons (to nearest 0.25 million))

From	To										
	United States	Canada	Other western hemisphere	Western Europe	Africa	South-East Asia	Japan	Australasia	Other eastern hemisphere	Destination not known	Total exports
United States	—	1.5	4	3.75	0.25	0.25	2	0.25	0.25	—	12.25
Canada	39.25	—	—	—	—	—	—	—	—	—	39.25
Caribbean	113.5	21	2	27.25	0.25	—	0.75	—	—	—	164.75
Other Western Hemisphere	1	—	—	—	—	—	—	—	—	—	1
Western Europe	7.5	—	—	—	2.25	—	0.5	—	1.25	4.5	16
Middle East	20	13.5	22.5	378.75	25.25	48.5	191.75	15.5	20.25	19.5	755.5
North Africa	4.5	—	4.25	158.25	0.25	—	1	—	10.5	7.0	185.75
West Africa	5.5	2.5	6.25	55.5	—	—	2	—	—	7.5	79.25
South-East Asia	6.75	—	0.5	0.25	0.25	—	31.25	3	—	—	42
U.S.S.R., Eastern Europe	0.25	—	7	43.75	3	—	1.5	—	0.5	—	56
Other Eastern Hemisphere	0.5	—	—	0.5	0.25	1.5	0.25	—	—	—	3
TOTAL imports	198.75	38.5	46.5	668	31.75	50.25	231	18.75	32.75	38.5	1,354.75

Adapted from National Academy of Science, *Petroleum in the Marine Environment*, Washington, D.C., 1975, 107 p.

Environmental fluxes

MAN-GENERATED

The estimates for the loss of petroleum, both crude oils and refined products, to the oceans have been made by a number of groups. The recent assessment by the United States National Academy of Sciences Workshop (NAS, 1975) will be reviewed here. Their results are presented in Table 35.

TABLE 35. Estimate of petroleum hydrocarbons (crude oils and refined products) entering the oceans (in millions of tons per year)

Source	Petroleum flux
Natural seeps	0.6
Offshore production	0.08
Transportation	
LOT tankers	0.31
Non-LOT tankers	0.77
Dry docking	0.25
Terminal operations	0.003
Bilges bunkering	0.5
Tanker accidents	0.2
Non-tanker accidents	0.1
Coastal refineries	0.2
Atmosphere	0.6
Coastal municipal wastes	0.3
Coastal, Non-refining, industrial wastes	0.3
Urban runoff	0.3
River runoff	1.6
TOTAL	6.113

Adapted from National Academy of Science, *Petroleum in the Marine Environment*, Washington, D.C., 1975, 107 p.

Of the nearly 2.5 billion tons of oil produced in 1971, 1.355 billion tons (which included 1.1 billion tons of crude oil) were transported across the oceans. A total of 6,000 tankers were used to transport this material. An important release of oil to the sea results from the washing of the cargo tanks of tankers with sea water and the discharging of these washings overboard. To minimize this loss, a procedure known as Load-on-Top (LOT) has been introduced. Here washings are retained on board the ship and the oil is allowed to coalesce in a holding tank. The oil can then be incorporated into the next shipment. About 80 per cent of the tankers use LOT. If the technique were not in use, an annual loss of oil is estimated to be 4.25 million tons. Present losses by ships not using LOT are probably about 0.77 million tons per year. The losses by LOT ships depend upon the efficiency and concern of individuals involved in the transport. They are estimated to be 0.31 million tons per year.

Dry-docking of tankers for maintenance and refurbishing requires a rigorous cleaning of all tanks. About 50 per cent of the ships enter the dry-docks with tanks which have been cleaned at sea with a consequent discharge estimated to be about 0.25 million tons per year. The washings at the dry-dock are presumably not discharged into the oceans.

Spillage rates at terminal operations at well-controlled ports are of the order of 0.00011 to 0.00022 per cent of the amount pumped for large tankers. For the many ports handling smaller shipments, the losses most probably are significantly higher. NAS (1975) considers the range of losses at terminals to be between 0.0015 and 0.005 million tons per year with a preferred value of 0.003 million tons per year.

The rate of loss due to bilges and bunkering is estimated to be 0.5 million tons per year which is equivalent to about 10 tons per ship per year.

The variations in losses from tanker accidents over a period of a year depend to a great extent on the number of major ones. Estimates range between 0.05 and 0.25 million tons per year. The *Torrey Canyon* accident resulted in a loss of about

0.1 million tons. The NAS (1975) estimate is 0.2 million tons per year. Although there are about eight times as many non-tankers as tankers, their losses of oil to the sea are probably much smaller over-all. Their average size is smaller and they carry less oil, primarily as fuel. NAS (1975) estimates their petroleum loss to be 0.1 million tons per year with a possible range between 0.02 and 0.15 million tons per year.

The amount of oil introduced into coastal waters in the water effluent from seaboard refineries is estimated to be 0.2 million tons per year (NAS, 1975). Previous estimates were based upon a single cycle of coolant water which gave higher results. Recirculated water cooling or air cooling is much more common today than the single water pass through, and the above result is presumed to reflect the practical situation.

The atmospheric flux of petroleum derived hydrocarbons is estimated to be 68 million tons per year (NAS, 1975) and a part of this does enter the ocean system through washout with rain, through fallout, or through air-sea interactions. Two-thirds of this amount, or 45 million tons, per year enter from vehicles involved in transportation such as automobiles and planes. The remainder originates in fuel combustion in stationary sources, industrial processes and solvent and gasoline evaporation. The NAS (1975) estimate of atmospheric flux is based upon a division of the volatilized hydrocarbons into two groups: a reactive component (33 million tons per year) which is involved in photochemical oxidative reactions and a non-reactive component (35 million tons per year). The assumption is made that all of the reactive fraction and 10 per cent of the non-reactive fraction is altered in the atmosphere and does not return to the earth's surface as hydrocarbons. Thus, 4 million tons per year of hydrocarbons are postulated to return to the earth's surface. Between 10 and 20 per cent of this returned hydrocarbon is estimated to fall upon the ocean surface (between 0.4 and 0.8 million tons per year) based upon relative areas of land and sea, on global precipitation patterns and the distribution of atmospheric particulates over land and sea areas. Thus, an average value of 0.6 million tons per year of petroleum hydrocarbons via atmospheric fallout is proposed.

The flux of oil to the ocean from offshore drilling and production operations is estimated to be 0.08 million tons per year (NAS, 1975). Three-quarters, or 0.06 million tons per year, is lost from major accidents and spills, usually of an unpredictable nature. Such spills are taken to be greater than 50 barrels each. Normal operations are presumed to account for 0.02 megatons per year through discharges of 50 barrels or less. The data were obtained from analysis of United States operations on the Gulf of Mexico coast and have been extrapolated to a global basis taking into account differences in operating procedures in countries other than the United States.

The amounts of petroleum discharged in sewage wastes to the oceans from coastal cities can be approached with a knowledge of the *per capita* contributions of oil and grease. Estimates range from 6 to 27 grammes per person per day and NAS (1975) accepted as a reasonable value 16 grammes per person per day. The average contribution for the southern California metropolitan complex (11 million inhabitants) and from the Island of Oahu (1 million inhabitants) was about this value. For Rio de Janeiro with a population of 3 million people a range of 9 to 26 grammes per person per day was observed.

Investigations in southern California indicated that one-half of the oil and grease in sewage originates from petroleum. Thus the sewage flux of 8 grammes *per capita* per day was found. About 50 per cent of this comes from municipal waste waters (which collect discharged materials from

gas stations, garages, commercial operations, etc.) and the remaining 4 grammes
per person per day are carried by flows
of industrial wastes discharged to municipal sewage systems. In the United States,
about 38 million people live in the coastal
zone. Thus the total flux from the United
States is 0.1 million tons per year. Extrapo-
lating to a world-wide basis on the assumption that since the United States uses one-
third of the petroleum, it provides one-third
of the sewage wastes, a global flux of
0.3 million tons per year is found.

NAS (1975) points out that inputs from
industry (other than from refineries and
those entering municipal waste systems)
appear to be about the same as that from
domestic waste waters. Thus, coastal-zone
industrial outfalls are taken to introduce
about 0.3 million tons per year to the
oceans.

Another contribution to municipal
waste is the 'urban runoff' which is the flow
of petroleum hydrocarbons deposited in
urban areas from oil-heating systems, atmospheric fallout, and operation of automobiles, into storm drains and thence to
receiving waters. Oil and grease in storm
drainage appear to be about one-third of
that in municipal and industrial waste
waters from the same area in studies
made in southern California, New York and
Sweden. NAS (1975) assumes that 75 per
cent of the oil and grease in urban runoff is of petroleum origin, as compared to
50 per cent in municipal and industrial
waste waters. Using these assumptions, a
best estimate of world input of petroleum
hydrocarbons from urban storm drainage is
0.3 million tons per year with a probable
range of 0.1 to 0.5 million tons per year.

NAS (1975) submits that previous estimates of the flux of river-borne petroleum
hydrocarbons to the sea have been low
inasmuch as the hydrocarbons associated
with the silt load have been ignored. Previous calculations have assumed a mean oil

concentration of 0.085 mg/l. However, recent measurements indicate that silt in the
Mississippi river carries a petroleum hydrocarbon load of 0.3 mg/l. Thus, a world input
of 1.6 million tons per year is found for
river-derived petroleum hydrocarbons.

Over-all then, a total of 5,750,000 tons
per year of petroleum hydrocarbons, both
refined products and crude oil, are mobilized to the ocean system by man. These
chronic and increasing losses can be compared with those that took place during
the Second World War when a substantial
amount of oil products was introduced into
the oceans through the sinking of ships.
Revelle *et al.* (1971) indicate that there
were ninety-eight United States-controlled
tankers sunk or damaged, each containing
around 10,000 tons of oil, by military activities not involving submarines. This injection of about 1 million tons is increased
by an estimated 3 million tons of oil lost
through the sinking of tankers by submarines. Altogether more than 4 million tons
were lost or a little less than the present
annual input through man's activities.

NATURAL SEEPAGE

That portion of the oil present in the marine system resulting from the direct discharge of submarine seeps into the water
provides the baseline upon which the man-
mobilized petroleum hydrocarbons are assessed. Wilson *et al.* (1974), using the following arguments, have estimated the total
world-wide annual marine seepage to be
0.6 million tons per year (probable range
is 0.2 to 6.0 million tons per year), perhaps
10 per cent of the anthropogenic flux.

Geological and geochemical conditions
apparently govern where seeps do and
do not occur. Seeps originate from both
commercial and non-commercial oil reservoirs. There appears to be a strong correlation between areas of high seepage
and current tectonic activity. For example,

areas of high seepage in southern California and southern Alaska are deformed, and have had recent and extensive earthquake activity. Ghawar, the world's largest oil field, is located on the Arabian platform, and shows no recent earthquake activity. It has little or no seepage.

Geologic criteria were developed for evaluating the seepage potential of offshore areas, and three categories were defined: (a) areas with a high potential for seepage characterized by a high incidence of earthquakes, strike-slip faulting and thick geochemically mature, tertiary sediments; (b) areas with a moderate potential for seepage, characterized by a low incidence of earthquakes, diapiric or intrusive structures and growth faulting associated with giant river-fed submarine fans; and (c) areas with a low potential for seepage, characterized by little or no earthquake activity, with older sediments or geochemically immature young sediments and little indication of recent deformation. Areas of the world prone to seepage, as determined by the above criteria, are given in Table 36. On the basis of measured seepages in a number of areas, flow rates of 100 barrels per day per 1,000 square miles were used

TABLE 36. Seepage-prone areas of the oceans

Ocean	Number of 1,000-square-mile areas of:		
	High-potential seepage	Moderate-potential seepage	Low-potential seepage
Pacific	568	2,715	1,241
Atlantic	381	3,030	3,289
Indian	145	2,318	880
Arctic		1,648	718
Southern		142	134
TOTAL	1,094	9,853	6,262

Reproduced from R. D. Wilson, P. H. Monaghan, A. Osanik, L. C. Price and M. A. Rogers, 'Natural Marine Oil Seepage', *Science*, Vol. 184, 1974, p. 857–65.

for the high potential seepage areas, 3 barrels per day per 1,000 square miles for areas of moderate seepage and 0.1 barrels per day per 1,000 square miles for areas of low seepage. The value for moderate seepage areas is the geometric mean of the high and low rates. Combining the seepage-prone areas of the world with the estimated flow rates, it is found that areas of high-potential seepage contribute about 45 per cent of the world-wide seepage, areas of moderate seepage about 55 per cent and areas of low seepage about 1 per cent. The Pacific Ocean has a larger amount of high-potential seepage areas than do the Atlantic, Indian, Arctic and Southern oceans.

Environmental fates

Spilled petroleum, following release to the oceans, spreads over surface waters. The extent of spreading depends upon the nature of the material and the prevailing wind and current systems. At this time physical, chemical and biological processes begin which alter the composition of the oil which, because of its immiscibility with water, is initially present as a separate phase. Although the rates and degrees at which changes take place in the environment are not known in detail, the general directions can be predicted with some degree of certainty.

Domestic or industrial oil wastes may be introduced as particles, emulsions or dissolved substances. Their subsequent fate depends in part upon this initial state of entry. These hydrocarbons may be subject to any of the processes listed below.

EVAPORATION

There is a rapid loss of low molecular weight hydrocarbon molecules (12 or less carbon atoms) from the surface oil slick. Since C_4 to C_{12} hydrocarbons make up nearly 50 per cent of an average crude oil, there may be a very rapid substantial loss of oil to the atmosphere after a spill. The rougher the sea state, the more rapid will be the evaporation process. The formation of oil globules and the bursting of bubbles creates greater surface areas of petroleum at which the vaporization process can occur. The fate of these low molecular weight hydrocarbons in the atmospheric reservoir, especially with regard to their potential return to the oceans, is not known.

SOLUTION

The lower its molecular weight and the greater its polarity, the more soluble a hydrocarbon will be in water. For the normal paraffin hydrocarbons, the approximate distilled water solubilities have the following values in parts per million: C_5, 400; C_6, 10; C_7, 3; C_8, 1; C_{12}, 0.01; C_{30}, 0.002. For benzene, toluene and the xylenes, the solubilities are 1,800 p.p.m., 500 p.p.m. and 175 p.p.m., respectively. The solubilities in sea water are estimated to be 25 per cent less than these values due to the salting-out effect, if the reduction is similar to that for inert gases such as the noble gases and nitrogen.

Inorganic and microbiological degradation processes can produce more soluble compounds. For example, the oxidation of *n*-octanol (solubility 1 p.p.m.) yields *n*-octanic acid (solubility 600 p.p.m.).

EMULSIFICATION

Both water-in-oil and oil-in-water emulsions can form, assisted by the presence of surface active constituents in ocean waters. For water-in-oil emulsions, the water uptake is a function of the type of oil: Kuwait crude takes up 50 per cent water in two to four hours while a Tia Juana crude takes two days to absorb 3 per cent water (NAS, 1975). The water-in-oil emulsion known as

the 'chocolate mousse' has been suggested as the precursor for the beach tars and tar lumps, following dissolution and chemical alteration of its components.

There is little information about the oil-in-water emulsions.

SEDIMENTATION

Sedimentation will take place when the oil components become aggregated into tarry lumps, emulsions or residues whose densities exceed that of sea water. Aging of petroleum through evaporation and solution of the low molecular weight compounds and the uptake of mineral matter on to its surfaces increases its density. The initial product of evaporation and solution, combined with the other processes of oxidation and degradation, are the tar balls, the semi-solid globules which float at the surface. Further degradation of these tar particles leads to the formation of smaller, denser forms that sink to the sea floor (Morris and Butler, 1973).

Some crude oils and the heaviest distillates such as Bunker C have specific gravities very close to 1. Thus, only small amounts of mineral matter associated with them are needed to bring about the depositional process. In shallow coastal waters, turbulent motion can bring surface sedimentary material into suspension. Subsequent contact with oil particles can result in their adherence and a sinking of the oil/sediment masses.

OXIDATION

Exposure of petroleum constituents to oxygen and light in surface waters can result in their oxidation. The rate of oxidation depends upon the particular substances within the product. For example, alkyl-substituted naphthenes will be oxidized more rapidly than normal alkanes. The surface slicks, in contrast to emulsions or tar balls, will be more readily subject to destruction by photo-oxidative processes. NAS (1975) indicates that in laboratory experiments with light sources similar in spectrum and intensity to sunlight, a slick 2.5 µm thick was decomposed in 100 hours.

MICROBIAL DEGRADATION

There is a general assumption pervading the older marine literature that petroleum is entirely biodegradable and that bacteria in the sea are primarily responsible for its decomposition. This concept has come in for critical examination and appears to be wanting (Floodgate, 1972). The large variety of laboratory experiments conducted upon various types of oils has produced a situation in which the results are difficult to apply to the natural environment. Further, extrapolation of laboratory data to an interpretation of environmental events can be misleading. What we can say at present is that marine micro-organisms are capable of degrading some components of oils, but that the rates of decomposition in the natural situation are poorly known.

The estimate of Zobell (1964) for the degradation rate of petroleum, 36–350 g/m^3 per year, is three orders of magnitude higher than that measured *in situ* for *n*-dodecane by Robertson *et al.* (1973). This is an easily oxidizable hydrocarbon.

Increasing temperature results in increasing biodegradation. The availability of nutrients, nitrogen and phosphorus can determine microbiological activity. In surface waters of the open ocean, where these elements can be in very low supply, the rates of oil breakdown by micro-organisms can be limited. NAS (1975) suggests other factors that may limit biodegradation: insufficient organic substrate for the development of viable colonies of micro-organisms; the presence of alternative carbon sources; and the presence of microbial predators.

ENTRY INTO THE MARINE BIOSPHERE

The entry of petroleum hydrocarbons into the marine food web is clearly demonstrated by the many observations of oil burdens in a variety of organisms (see above). Petroleum is available in the form of dissolved phases, of dispersions, in particles such as the floating tar lumps or as sorbed phases to the particles of sea water. In addition, petroleum can be incorporated in an organism through the food it eats. Once in an organism it can be metabolized, stored or excreted. There are some studies on the metabolism and alteration of petroleum hydrocarbons in marine fish and invertebrates. Aromatic and paraffinic hydrocarbons are degraded by marine fish and some marine invertebrates. Other marine invertebrates, phytoplankton and some zooplankton were unable to break down such compounds over a period of a month (NAS, 1975). Several species of copepods are reported to be able to degrade paraffinic hydrocarbons.

FATE IN THE SEDIMENTS

Detailed studies of a September 1969 spill of fuel oil in Buzzards Bay, Massachusetts, indicated that petroleum hydrocarbons have a considerable persistence in marine sediments (Blumer and Sass, 1972b).

Normal alkanes persisted for at least two years in the sediments. The rates of decomposition for the branched and cyclic hydrocarbons were slower than for their normal counterparts. For periods of two years after introduction of the oil, the isoprenoids (phytane, pristane and the C$_{18}$ homologue) and alicyclic and aromatic hydrocarbons were easily measurable in the deposits.

The disappearance of the oils from the sediments resulted from both microbial degradation and through solution, with the latter being quantitatively more important. This was evident from the relative increase between March 1970 and April 1971 of the more highly substituted benzenes, naphthalenes and tetrahydronaphthalenes at the expense of the more soluble lower homologues such as naphthalene and the C_1 to C_3 alkylnaphthalenes.

Natural petroleum hydrocarbons are preserved in the sedimentary column for geological time periods and the oil introduced to the deposits by man may remain for similar lengths of time.

Petroleum introduced to intertidal and beach deposits is subject to a variety of processes resulting in its dispersion and/or degradation. In rocky areas, it accumulates in pores and on algal and animal growths, and its removal by tidal flushing can take place in periods of months to years. Accumulation in intertidal and beach deposits is usually a function of grain size, with the fine-grained clay minerals absorbing greater quantities than the silts or sands. Through reworking of these sedimentary phases by tidal activity there is lateral spreading with a greater possibility of oxidation and dissolution. Oil or tar globules on beaches tend to take up sediment particles on their surfaces. Upon loss of the lighter components through volatilization or solution, they behave as solids.

Environmental levels

The difficulties in analytically characterizing petroleum hydrocarbons has resulted in a dearth of determinations of environmental levels. There are few published analyses and the few examples are more illustrative of the concentrations in a given region at a given time rather than of a more general nature. Man-mobilized, biogenic and seepage hydrocarbons are usually not distinguished.

SEA WATER

Three domains of petroleum may be categorized for the waters of the ocean: the dissolved and particulate phases in the water column; the sea surface microlayer; and the floating tar balls.

Determination of dissolved petroleum hydrocarbons are scanty. IDOE (1972) indicates that water off the Louisiana coast had concentrations of 0.63 µg/l for hydrocarbons in the C_{16} to C_{34} range and 0.03 and 30 µg/l were measured for C_1 to C_3 hydrocarbons for two samples of Gulf of Mexico water.

One deep-sea profile of extractable organics and hydrocarbons from a station near Bermuda is reported by NAS (1975) from the unpublished data of R. A. Brown (Fig. 22). The highest concentrations of hydrocarbons in sea water are in the upper ten metres with levels of 6 µg/l. At greater depths in the upper one hundred metres (the mixed layer), the hydrocarbon concentration is about 3 µg/l. At 1,000 metres values of 1 µg/l are found decreasing to zero at depths greater than 2,000 metres. On the basis of this profile NAS (1975) estimated the oceanic burden of dissolved hydrocarbons to be 400 million tons.

Sea surface microlayer concentrations were reported to the NAS (1975) workshop from the University of Rhode Island laboratories. The enrichment in this zone, compared to the water lying below it was not large: at most a factor of 2.5.

Chloroform extracts of unpolluted Atlantic waters yielded a collective of hydrocarbons whose composition was similar to that of marine algae (Barbier *et al.*, 1973). N-paraffins occur to an extent of about 12 per cent from C_{14} to C_{37} with a maximum at C_{27} to C_{30} (Fig. 23). Coastal waters indicated a polluted character by the presence of lower molecular weight hydrocarbons C_{11} to C_{20} (Fig. 23). These authors indicate that the chloroform extracts contain all of the hydrocarbons and

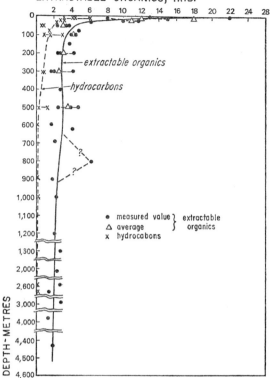

FIG. 22. Deep-sea profile of extractable organics and hydrocarbons from a station near Bermuda (32°18′ N.; 65°32′ W.). From NAS (1975).

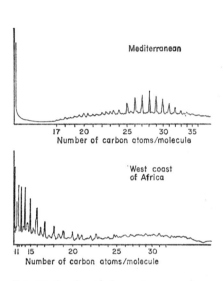

FIG. 23. Gas-liquid chromatograms of an unpolluted sample taken at 50 metres depth near Villefranche in the Mediterranean and of a polluted sample taken off the west coast of Africa from a depth of 50 metres (Barbier *et al.*, 1973).

represent between 10 and 20 per cent of the total organic matter for tropical waters and from 40 to 55 per cent for the Antarctic oceanic waters. The average concentration of hydrocarbons in the world ocean was estimated at 10 μg/l which would give a total burden of 14 billion tons, significantly higher than the NAS (1975) estimate, but many orders of magnitude larger than the estimated annual input of man-mobilized petroleum hydrocarbons.

The Mediterranean appears to be badly polluted with petroleum hydrocarbons as evidenced by analyses of its surface films, subsurface oil/water emulsions and zooplankton (Morris, 1974). Samples were obtained in September and October 1973, between the west coast of Crete and Cyprus. The surface films contained 40–230 mg/m² of organic materials where less than 5 per cent were natural products. The complex mixture included polyaromatic compounds, saturated alkanes (straight chained and branched) and isoprenoids (pristane and phytane).

Water/oil emulsions, below the surface

films, were found off the Libyan and Egyptian coasts, probably resulting from the cleaning of tanks of oil-carrying vessels. Net collections of the emulsions indicated levels of 100–500 g/km^2 near the coast with lower values at greater distances from land. The emulsions contained over 75 per cent pollutant hydrocarbons, and the mixtures were similar in composition to those of the surface films.

Tar balls had densities of 0.7–10 kg/km^2 in areas where no heavy oil slicks were present. Zooplankton collected from the surface waters contained hydrocarbons similar in composition to those of the emulsions and surface films.

An empirical relationship to differentiate between open-ocean clean water and water contaminated by man-mobilized low molecular weight hydrocarbons has been proposed by Swinnerton and Lamontagne (1974). The Contamination Index (CI) is defined by $CI = 1/3(C_1/C_1^* + C_2/C_2^* + C_3/C_3^*)$,

where C_1, C_2 and C_3 are concentrations of methane, ethane and propane in the water under consideration and C_1^*, C_2^* and C_3^* are the average baseline concentrations of these hydrocarbons in clean ocean water. On the basis of 427 clean ocean water samples, C_1^*, C_2^* and C_3^* were found to have values of 49.5, 0.50 and 0.34 nl/l, respectively.

The criterion for contaminated waters is a value of CI greater than 5. Possibly contaminated waters have values of CI between 3 and 5. Uncontaminated waters have CI values of less than 3. Highly contaminated waters were found in the Gulf of Mexico, while the Caribbean Sea was clean in its surface waters on this basis.

The most extensive set of analyses in the waters involve the tar balls; these analyses are summarized in Table 37. There is a 'patchiness' in the distribution of these materials and single measurements are not very meaningful. NAS (1975) presents a case for a strong co-variance between ob-

TABLE 37. Tar densities in the world oceans

Location	Area (10^{12} m^2)	Tar (mg/m^2) Maximum	Tar (mg/m^2) Mean	Total tar (10^3 tons)
North-West Atlantic Marginal Sea	2	2.4	1	2
East Coast Continental Shelf	1	10	0.2	0.2
Caribbean	2	1.2	0.6	1.2
Gulf of Mexico	2	3.5	0.8	1.6
Gulf Stream	8	10	2.2	18
Sargasso Sea	7	40	10	70
Canary and North Equatorial Current	3	1,000?	?	? (large)
Mediterranean	2.5	540	20	50
Indian Ocean	75	?	?	? (large)
South-West Pacific	45		<0.01	<0.5
South-East Pacific	45		?	?
Kuroshio System	10	14	3.8	38
North-East Pacific	40	3	0.4	16

Reproduced from National Academy of Science, *Petroleum in the Marine Environment*, Washington, D.C., 1975, 107 p.

TABLE 38. Relation between occurrence of spilled oil and occurrence of tar concentration

Location	Area (10^{12} m^2)	Spilled petroleum (mg/m^2/yr)	Tar supply rate (mg/m^2/yr)	Tar found mean (mg/m^2)
North Atlantic	33	17.45	6.13	5
Mediterranean	2.5	108	38	20
Kuroshio System	10	33	11.6	3.8
North-East Pacific	40	0.74	0.26	0.4
South-West Pacific	45	<0.05	<0.02	<0.005

Reproduced from National Academy of Science, *Petroleum in the Marine Environment*, Washington, D.C., 1975, 107 p.

served high levels of oceanic tars and known tanker routes of high traffic (Table 38). The tar flux to a given area is assumed to represent 35 per cent of the oil spilled, assuming no inputs from oil seeps. There are as yet no measurements of the abundance of tar lumps in areas of known natural seepage.

SEDIMENTS

The hydrocarbon concentrations in marine sediments show variations over four orders of magnitude (NAS, 1975), see Table 39. The unpolluted coastal areas and open-ocean sediments contain 1–4 p.p.m. dry weight of hydrocarbons; less than 100 p.p.m. in coastal sediments that are unpolluted; and up to 12 parts per thousand in highly polluted areas. In most of these analyses it is difficult to distinguish between hydrocarbons of a biological origin and the petroleum hydrocarbons dispersed by man. Blumer and Sass (1972b), however, indicated a technique by which such a distinction might be made. By following the composition of sediments as a function of depth in the Buzzards Bay oil zone, they were able to show that materials extracted at depths greater than 7.5 centimetres contained little or no fuel-derived hydrocarbons, while the surface levels of the

deposits contained substantial quantities of such oils.

The biological hydrocarbons could be distinguished through a marked predominance of normal paraffins with an odd number of carbon atoms, especially those greater than C_{21}. Except for some very young oils, petroleums show no odd-even preferences in their normal paraffin contents. The surface sediments contained hydrocarbons boiling within the temperature range of fuel oils, while those at greater depths contained predominantly hydrocarbons boiling above the fuel oil range. In addition, pristane, the principal resolved component in gas chromatographic records of partially degraded fuel oils, was only found in very minor amounts in the deep sections, while it is a major component in the surface layers along with the adjacent C_{18} and C_{20} isoprenoids. Hydrocarbon separates from sediments containing fuel oils show an unresolved envelope in their gas chromatographic patterns for the fuel oil boiling range, a characteristic not found for separates from sediments where there is no other evidence for fuel oils. Finally in their gas chromatographic assays of hydrocarbons, the upper strata of sediments yielded the 'fingerprint peaks' corresponding to specific components in fuel oil contaminants. These techniques could be applied to other coastal

Table 39. Hydrocarbons in sediments

Location	Sample depth	Water depth (metres)	Concentration (p.p.m., dry weight)	No. of samples
Buzzard's Bay, Massachusetts, Wild Harbor river (over 2½-year period)	Top 10 cm, sand and clay	<3	250–1,650	~180
Buzzard's Bay, Massachusetts, Station 37, subtidal unpolluted	Top 10 cm	11	38–70	10–20
Buzzard's Bay, Massachusetts, Silver Beach (over 2½-year period)	Top 10 cm sand and clay	<2	500–12,000	~60
Mississippi coastal bog	—	0–1.5	350	1
Narragansett Bay, Rhode Island, mouth of Providence river (head of bay)	Top 8–10 cm	?	820–3,560	2
Middle of bay	Top 8–10 cm	?	350–440	2
Mouth of bay	Top 8–10 cm	?	50–60	2
Vineyard Sound, Massachusetts	Top 7 cm	6–9	1.7 *n*-alkanes only	1
Chedabucto Bay, Nova Scotia (over 2-year period)	Top 5 cm	3	34–420	7
Chedabucto Bay, Nova Scotia (over 2-year period)	Top 5 cm	12	11–1,240	7
Coast of France, Le Havre vicinity	'Surface'	Subtidal	450	1
Seine estuary	'Surface'	Subtidal	33	1
Coast of France, Le Havre vicinity (different from the one above)	'Surface'	Subtidal	920	1
Bay of Veys	'Surface'	Subtidal	38	1
Port Valdez, Alaska	'Surface'	Subtidal	0.5–2.5 (C_{16}–C_{28} only)	?
Gulf of Batabano, Cuba	?	?	15–85	10
Orinoco Delta, Venezuela	?	?	27–110	10
Gulf of Mexico, open ocean	?	?	12–63	10
Mediterranean, open ocean	?	?	29	1
Cariaco Trench	?	?	56–352	16
South-east Bermuda to base of rise near Hudson Canyon	Top 5 cm of sediment	>3,000	1–4	5

Adapted from National Academy of Science, *Petroleum in the Marine Environment*, Washington, D.C., 1975, 107 p.

ocean deposits to ascertain the present fluxes of man-mobilized petroleum hydrocarbons.

BIOSPHERE

Perhaps no part of the ocean has been free from the entry of man-mobilized petroleum and consequently no marine organisms are devoid of such pollutants. NAS (1975) collated analyses of hydrocarbons in marine organisms taken from various types of waters (Table 40). The compilers of the table 'attempted to exclude biogenic hydrocarbons by subtracting levels reported for control samples (i.e. unpolluted) in those cases where the authors have not already done so'. Even for samples taken from

TABLE 40. Petroleum
hydrocarbon levels in
marine macro-organisms

Organisms	Area type*	Hydrocarbon type	Estimated HC amount (μg/g)
Macroalgae			
Fucus	4	Bunker C**	40 dry
Enteromorpha	4	No. 2 fuel oil	429 wet
Sargassum	1	C_{14-30} range	1–5 wet
Higher plants			
Spartina	4	No. 2 fuel oil	15 wet
Molluscs			
Modiolus, mussel	4	No. 2 fuel oil	218 wet
Mytilus, mussel	4	No. 2 fuel oil**	36 dry
Mytilus	4	Bunker C**	10 dry
Mytilus	4	Bunker C, aromatics	74–100 wet
Mytilus	3	$n\text{-}C_{14-37}$ **	9 dry
Mya, clam	4	No. 2 fuel oil	26 wet
Pecten, scallop	4	No. 2 fuel oil	7 wet
Littorina, snail	4	Bunker C, aromatics	46–220 wet
Mercenaria, clam	3	C_{16-32} range	160 dry
Crassostrea, oyster	2	Polycyclic aromatics	1 wet
Crustacea			
Hemigrapsus, crab	4	Bunker C**	8 dry
Mitella, barnacle	4	Bunker C**	8 dry
Lady crab	3	C_{14-30}	4 wet
Plankton	2	Benzopyrene	0.4 wet
Sargassum shrimp	1	C_{14-30}	3 wet
Lepas, barnacle	1	C_{14-30}	6 wet
Portunus, crab	1	C_{14-30}	34 wet
Planes, crab	1	C_{14-30}	11 wet
Fish			
Fundulus, minnow	4	No. 2 fuel oil	75 wet
Anguilla liver, eel	4	No. 2 fuel oil	85 wet
Smelt	3	Benzopyrene	0–5 dry
Flatfish	2	C_{14-20}	4 wet
Flying fish	1	C_{14-20}	0.3 wet
Sargassum fish	1	C_{14-20}	1.6 wet
Pipefish	1	C_{14-20}	8.8 wet
Triggerfish	1	C_{14-20}	1.7 wet
Birds			
Herring gull, muscle	4	No. 2 fuel oil	535 wet
Echinoderm			
Asterias, starfish	4	Bunker C, aromatics	20–147 wet
Luidia, starfish	2	C_{14-30}	3.5 wet

* 1, oceanic; 2, chronic pollution, coastal; 3, chronic pollution, harbour; 4, single spill.
** *n*-alkanes only.
Reproduced from National Academy of Science, *Petroleum in the Marine Environment*,
Washington, D.C., 1975, 107 p.

the open ocean, there appears to be parts per million levels of anthropogenic hydrocarbons in the organisms.

There is the possibility that plankton samples are contaminated by slick-borne hydrocarbons or tar balls during their collection with nets.

ATMOSPHERE

There are few measurements of petroleum hydrocarbons in marine airs. Rasmussen (cited in NAS, 1975) found 30 $\mu g/m^3$ of C_3 to C_{12} hydrocarbons in tradewinds from the north-east side (upwind) of the island of Hawaii. Quinn and Wade (cited in NAS, 1975) have shown that 98 per cent of the C_{14} to C_{33} hydrocarbons are in the gas phase or as fine particulate (particles smaller than 0.02 μm) in samples taken from a tower on the west coast of Bermuda. The vapour phase contained 5 $\mu g/m^3$ (30 per cent resolved *n*-alkanes and branched alkanes: 70 per cent unresolved aromatics and naphthenics). The particulate phases contained 0.1 $\mu g/m^3$. These concentrations are much lower than the hydrocarbon concentrations found over land of 50 to 100 $\mu g/m^3$ (Rasmussen, cited in NAS, 1975).

IMPACTS ON LIVING SYSTEMS

The entry of man-generated petroleum hydrocarbons into the marine food web is clearly demonstrated by the many recent analyses of a variety of organisms (Table 40). Concentrations of petroleum hydrocarbons in the organisms are in the parts per million range in contrast to parts per billion levels in sea water. The effects of such low body concentrations on the health of the organisms, if any, have not as yet been determined.

Gordon and Prouse (cited in NAS, 1975) found that No. 6 fuel oil stimulated photosynthesis in a mixed natural population of phytoplankton at sea-water concentrations of 10 to 30 p.p.b., but repressed it at concentrations between 60 and 200 p.p.b.

There are other laboratory investigations describing the effect of a specific petroleum product or crude oil upon an organism, but their applications to the chronic backgrounds of oil that organisms are exposed to are difficult to make. The stress of a laboratory environment may bring an organism near its limits of tolerance of any one of a number of chemical or physical factors, such as temperature, oxygen content of the water, etc. The added stress of a petroleum pollutant may bring about a response that would not occur for the same concentration of the pollutant in a natural situation.

Besides the direct effect of petroleum hydrocarbons upon the metabolism of organisms, there are other levels of interactions that must be considered. For example, certain petroleum components interfere with the processes of chemoreception through the blocking of the detection organs. An individual animal so affected may be placed at a competitive disadvantage in its search for food (IDOE, 1972). Reproductive processes can be hampered by the masking of the presence of pheromones (chemicals produced by organisms to attract the opposite sex). In the case of the lobster, exudations of pheromones by the female are necessary to induce copulation. Pheromone activity in other marine organisms is still poorly studied. The 1972 IDOE report also suggests that hydrocarbons may inhibit reproduction either through reduction in the viability of the gametes or by killing early life stages of animals.

Experiences with oil spills have been much more revealing as to the vulnerability of the marine environment to petroleum. In spite of variations in the composition of the oil, the nature of the marine environment in which the spill took place, and the amounts of oil, certain generalizations can be made. Mass mortalities of organisms have been observed in all major accidents. Pelagic diving birds are especially vulnerable. They catch their food by diving into the sea, and, in general, they are weak fliers. Upon diving through a surface coated with oil, they become coated. During preening, they ingest the oil, which apparently is partially responsible for their deaths.

On the other hand, fish often escape the toxicity of massive amounts of petroleum by swimming away from the area that became polluted. The members of the benthic community, being less mobile, can suffer large numbers of mortalities as the oil sinks to the sediments, as was the case in the West Falmouth Spill (Blumer and Sass, 1972a). The exposures in the sediments and waters continued for at least two years after the accident in the more highly polluted areas.

But what of the impact upon fisheries and man? NAS (1975) points out that in the highly oil-polluted areas of the Gulf of Mexico near Louisiana where many oil-field operations have been unintentionally spilling materials, commercial fishing is in a most healthy state. Tainting of oysters with oil (levels sometimes reaching 500 p.p.m.) makes them unmarketable because of bad taste, unless the oil is removed by taking them to unpolluted areas for several months. Other activities such as dredging and the resultant silting effects from the laying of oil lines and the construction of rigs has caused damage to the coastal ecosystem and fisheries (NAS, 1975).

Finally, there are possible effects upon human health through the ingestion of fish or shellfish contaminated with petroleum, even though the levels are very low. The concern involves the ingestion of carcinogenic compounds, especially the polycyclic aromatic hydro carbons, present in petroleums. Clearly, the consumption of shellfish can add small amounts of such substances to those already taken in through inhalation of cigarette smoke and of the products of fossil fuel combustion. The whole area of concern needs a systematic investigation to remove the clouds of uncertainty surrounding this subject.

Bibliography

BARBIER, M.; JOLY, D.; SALIOT, A.; TOURRES, D. 1973. Hydrocarbons from sea water. *Deep-sea research*, vol. 20, p. 305–14.

BLUMER, M.; SASS, J. 1972a. Oil pollution: persistence and degradation of spilled fuel oil. *Science*, vol. 167, p. 1120–2.

——. 1972b. Indigenous and petroleum derived hydrocarbons in a polluted sediment. *Mar. Pollut. Bull.*, vol. 3, p. 92–3.

FLOODGATE, G. D. 1972. Microbial degradation of oil. *Mar. Pollut. Bull.*, vol. 3, p. 41–3.

IDOE. 1972. *Baseline studies of pollutants in the marine environment and research recommendations. Deliberations of the International Decade of Ocean Exploration, Baseline Conference May 24–26, 1972, New York.*

MORRIS, R. J. 1974. Lipid composition of surface films and zooplankton from the eastern Mediterranean. *Mar. Pollut. Bull.*, vol. 5, p. 105–9.

MORRIS, B. F.; BUTLER, J. N. 1973. Petroleum residues in the Sargasso Sea and on Bermuda beaches. *Proc. Joint Conference on the Prevention and Control of Oil Spills*, p. 521–30.

NAS. 1975. *Petroleum in the marine environment.* Washington, D.C., National Academy of Science, 107 p.

REVELLE, R.; WENK, E.; KETCHUM, B. H.; CORINO, E. R. 1971. Ocean pollution by petroleum hydrocarbons. In: W. H. MATTHEWS, F. E. SMITH and E. D. GOLDBERG (eds.), *Man's impact on terrestrial and oceanic ecosystems*, p. 297–318. Cambridge, Mass., MIT Press.

ROBERTSON, B.; ARHELGER, S.; KINNEY, P. J.; BUTTON, D. K. 1973. Hydrocarbon biodegradation in Alaskan waters. In: D. G. AHEARN and S. P. MEYERS (eds.), *The microbial degradation of oil pollutants*, p. 171–84. Baton Rouge, Louisiana State University.

SIMONOV, A. J.; JUSTCHAK, A. A. 1972. *Recent hydrochemistry variations in the Baltic Sea, 1st Soviet-Swedish Symposium on the Pollution of the Baltic*, p. 21–2. (Ambio special report No. 1.)

SWINNERTON, J. W.; LAMONTAGNE, R. A. 1974. Oceanic distribution of low-molecular-weight hydrocarbon baseline measurements. *Environ. Sci. Technol.*, vol. 8, p. 657–63.

WILSON, R. D.; MONAGHAN, P. H.; OSANIK, A.; PRICE, L. C.; ROGERS, M. A. 1974. Natural marine oil seepage. *Science*, vol. 184, p. 857–65.

ZOBELL, C. E. 1964. The occurrence, effects and fate of oil polluting the sea. *Proc. Intern. Conf. Water Pollution Res., London, 1962*, p. 85–109. Oxford, Pergamon Press.

7.   Litter

Introduction

The surface waters, the floor and the beaches of the world ocean are visibly being soiled by the discard of man-fabricated materials promiscuously released to the environment in the form of litter. Much of this rubbish has its origin in packaging materials—plastic, metal, cloth, glass or wood products.

An estimated flux of around 6.4 million tons per year of litter has been posited recently (NAS, 1975) and the contributions from various sources are given in Table 41.

Two types of litter were considered: domestic refuse, consisting mostly of packaging materials, and commercial refuse. Commercial refuse encompasses both cargo-associated wastes and lost fishing gear. About 75 per cent of the total is suspected of originating in losses during cargo hand-

ling from merchant shipping (Table 41).

The typical composition of non-cargo-associated wastes discharged from ships (less food and vegetable wastes) is: paper, 63 per cent; metal, 16.6 per cent; cloth, 9.6 per cent; glass, 9.6 per cent; plastics, 0.7 per cent; rubber, 0.5 per cent. Rapidly degradable materials such as vegetable and food wastes, as well as wrecks and materials which are placed to make artificial reefs, are not included in the NAS tabulations. Litter estimates were arrived at (a) by multiplying the number of people at sea engaged in a given activity by the average amount of trash that the activity generates and (b) by special methods for such sources as bulk cargo handling.

The presence of litter in seascapes creates an aesthetic insult, which appears to be the primary impact upon our society. Occasionally such objects as plastic sheets become caught in the water intake of ships, and ropes and wires have become entangled in the propellors of ships.

TABLE 41. Total litter estimates

Source	Amount of litter generated (10⁶ tons per year)
Passenger vessels	0.028
Merchant shipping	
Crew	0.110
Cargo	5.600
Recreational boating	0.103
Commercial fishing	
Crew	0.340
Gear	0.001
Military	0.074
Oil drilling and platforms	0.004
Catastrophes	0.100
TOTAL	6.360

Reproduced from National Academy of Science, *Petroleum in the Marine Environment*, Washington, D.C., 1975, 107 p.

Levels in the marine system

Litter accumulates at the boundaries of the ocean system—the ocean floor, the water surface, convergence zones, beaches, etc. Its distribution is highly uneven. Because of the great variety of products, the fate of litter must be analysed on a type by type basis, a study not yet undertaken.

The National Marine Fisheries Service of the United States estimated that there were 24,000 plastic items along a 60-mile stretch of beach in Amchitka, Alaska, remote from the activities of technological societies to the south. Among the litter were 12 tons of polypropylene gill nets and 7,000 gill-net floats. This material had washed up on to the beach during a six-month period in 1972.

There are a variety of litter measure-ments made upon both large and small objects. Venrick *et al.* (1973) during a survey of 12.5 km² over the central North Pacific cited fifty-three large man-made objects which included six plastic bottles, twenty-two plastic fragments, twelve glass fishing floats, four glass bottles, rope, an old balloon, finished wood, a shoebrush, a rubber sandal, a coffee can and three paper items.

The litter collected in nets dragged through marine waters primarily consists of plastic objects. One of the first studies noted the occurrence of polystyrene spherules, averaging 0.5 mm in diameter (with a range between 0.1 and 2.0 mm) in southern New England coastal waters (Carpenter *et al.*, 1972). The spherules were similar to those used for the fabrication of plastic-ware and probably originated from the polystyrene manufacturing plants of the eastern United States. The spherules were also found in the guts of fish from the area (USDC, 1973).

The most extensive study of the distribution and sources of plastics in the sea was made during July–August 1972 in the north-western Atlantic (Colton *et al.*, 1974). Samples were collected in neuston nets with a 0.945-mm mesh. The following categories of plastic particles were found:
1. White opaque polystyrene spherules with a mean diameter of 1.0 mm and a range of 0.2–1.7 mm. The source of these spherules is plastic manufacturing plants that discharge waste water into estuaries or rivers, which eventually transport the litter into the ocean.
2. Transparent polystyrene spherules containing gaseous voids with a mean diameter of 1.5 mm and a range of 0.9–2.5 mm. The sources and routes to the oceans are similar to the opaque spherules.
3. Opaque to translucent polyethylene cylinders or discs; mean diameter 3.4 mm; range in diameter, 1.7–4.9 mm; mean thickness, 2 mm; range in thickness,

1.1–3.4 mm. These materials, known as 'nibs' in the chemical trade, are used in fabricating plastic products, and the sources and routes to the ocean are similar to those of the polystyrene spherules.

4. Pieces of styrofoam, presumably from disposable cups.
5. Pieces of hard and soft, clear and opaque plastics, which appear to be parts of plastic containers, toys and so forth.
6. Sheets of thin, flexible wrapping material.

The plastic sheets and pieces were the most abundant and the most widely dispersed of these littering objects. The bulk of this material was 'disposable' wrapping and packaging material.

By the time such plastic particles reach the open ocean, they show signs of weathering by becoming brittle, presumably as a consequence of the loss of plasticizer (Carpenter and Smith, 1972). In addition, organisms attach themselves to the debris. In their samples from the Sargasso Sea, Carpenter and Smith noted hydroids and diatoms often attached to the surfaces.

What are the impacts of this litter upon biological activity? As noted above, it provides habitats for some organisms. Is it possible that these new ecological niches can enhance biological productivity?

The plastics themselves are apparently non-toxic. Spherules have been recovered from the stomachs of fish and birds. However, as yet no evidence of morbidity or mortality has been associated with this debris.

There are reports of deleterious effects none the less. The health of certain fish and marine birds has been disturbed by such commercial refuse as elastic bands and threads. Elastic bands have ringed sterlets in the Danube delta (Anon, 1971) and worked their way into their flesh. The bands produced mucus-covered ulcers and appeared to cause damage to the gills. Rubber thread cuttings are apparently mis-

taken for fish by puffins which swallow them (Parslow and Jefferies, 1972). Four of six puffins, which were being examined to ascertain their pollutant concentrations since they were found dead on beaches, contained strands of elastic thread in their alimentary tracts. It is unlikely, however, that the elastic was the cause of death. In this case evidence implicated oil pollution and in one case the striking of electrical wires. These threads of about 0.8 mm in cross section and 50–100 mm in length, often stranded, are used in the garment industry, probably discarded in the manufacturing process. The impact of the elastic bands on the health of the puffins is still undetermined.

The dumping of drums, military devices and other large artifacts into ocean waters constitutes a special and often serious litter problem. This is especially true where the materials chemically decompose and release noxious materials to the waters.

Bombs containing mustard gas $(ClCH_2CH_2)_2S$, from German ammunition depots were dumped following the Second World War into the deep waters of the Baltic. These have been implicated in a number of cases of acute food poisoning occasioned by eating contaminated fish eggs (Garner, 1973). A fishing vessel picked up one bomb and leakage from it contaminated the fish while the crew was cleaning them. Some crew members suffered skin burns from the contact. Other trawlers have hauled in such bombs and immediately returned them to the oceans. The amount of mustard gas bombs which has been dumped in the Baltic appears to be close to 20,000 tons.

Drums containing chemical wastes dumped into the North Sea are introducing a wide variety of synthetic organics to the waters after the containers rupture (Greve, 1971). These drums are disposed of not only in deep water but also in the shallow areas where they are often picked up by the nets of fishermen. The drums contain such wastes as lower chlorinated aliphatic compounds, vinyl esters, chlorinated aromatic amines and nitrocompounds and the pesticide endosulfan.

Overview

The littering of the marine environment, although not yet a priority problem in marine pollution, is more readily brought under control than the contamination of the ocean by man-generated chemicals. Since the primary inputs of litter come from ships, limitations upon the disposal of sewage and garbage both in coastal and in open-ocean waters would constitute a simple and effective remedy. The vessels of the United States are probably responsible for about one-third of the present flux of litter. Thus, regulation of that nation's vessel discharge would significantly reduce the litter burden of the world's oceans. In the face of an increasing input of litter, simple remedial and early preventive measures to maintain the integrity of the marine environment appear today most attractive.

Bibliography

ANON. 1971. Elastic band pollution. *Mar. Pollut. Bull.*, vol. 2, p. 1965.

CARPENTER, E. J.; ANDERSON, S. J.; HARVEY, G. R.; MIKLAS, H. P.; PECK, B. B. 1972. Polystyrene spherules in coastal waters. *Science*, vol. 178, p. 749–50.

CARPENTER, E. J.; SMITH Jr, K. L. 1972. Plastics on the Sargasso Sea surface. *Science*, vol. 175, p. 1240–1.

COLTON, J. B.; KNAPP, F. D.; BURNS, B. R. 1974. Plastic particles in surface waters of the northwestern Atlantic. *Science*, vol. 185, p. 491–7.

GARNER, F. 1973. Mustard oil on troubled waters. *Environment*, vol. 15, p. 4.

GREVE, P. A. 1971. Chemical wastes in the sea: new forms of marine pollution. *Science*, vol. 173, p. 1021–2.

NAS. 1975. *Assessing potential ocean pollutants*. Washington, D.C., National Academy of Science. 438 p.

PARSLOW, J. L. F.; JEFFERIES, D. J. 1972. Elastic thread pollution of puffins. *Mar. Pollut. Bull.*, vol. 3, p. 43–5.

UNITED STATES DEPARTMENT OF COMMERCE. 1973. Plastic wastes found on isolated Alaska beaches. *U.S. Department of Commerce News: March 29*. Washington, D.C.

VENRICK, E. L.; BACKMAN, T. W.; BARTRAM, W. C.; PLATT, C. J.; THORHILL, M. S.; YATES, R. E. 1973. Man-made objects on the surface of the central North Pacific Ocean. *Nature*, vol. 241, p. 271.

8.   The predictive mode

Two general types of problems are encountered in attempting to predict future exposure levels of pollutants that may jeopardize marine resources. The first involves the identification of materials that can in principle achieve unacceptable levels in our surroundings.

A description of the disposition of a pollutant in the marine environment can be very difficult to acquire, as the preceding chapters have illustrated. A more formidable set of problems becomes evident in going from the 'descriptive' to the 'predictive' mode.

What new materials, in the infancy of their production today, may jeopardize the resources of the oceans in the future? What marine areas will be especially vulnerable to the introduction of pollutants?

The screening of materials

Four characteristics of a substance[1] are useful as a guide in predicting potential hazards to the ocean system:

Rates and sites of release to the environment, obtained from production and use data.

Persistence in the environment.

Ability to accumulate in parts of the system or in organisms (bio-accumulation).

Levels of toxicity and the propensity for transforming into more or equally dangerous chemical forms.

1. A substance may vary with respect to its content of impurities. In some cases the latter may be of greater concern with respect to environmental impacts than the parent compound, for example, the toxic qualities of some PCBs have been attributed to the impurities, dioxan and benzofurans. In this connexion we are concerned with the impurities, not the parent compound.

Each of these factors is difficult to assess for a substance initially. Yet guidance can be obtained in a number of ways. In general, those substances with the highest production rates will attract scrutiny before those made in smaller amounts. As yet, we have no method to assemble production and use statistics on a world-wide basis for a potential pollutant. Still, indirect means, such as those used to ascertain the amounts of DDT at present being produced and the projections for the future (see Chapter 3), provide a basis for estimating potential hazards. However, sometimes even these numbers are unobtainable. An example would be those cases where the chemicals in question are generated in part as waste products (such as hexachlorobenzene—see Chapter 3). Consequently, rates of production and release can be quite difficult to obtain.

The persistence of a chemical species in the environment can sometimes be estimated through the extrapolation of results from laboratory experiments. Such may be the case with polychloroethylene (see Chapter 3, and below in this chapter). On the other hand, the rates of degradation of DDT in laboratory experiments are much more rapid than those observed in the field (Chapter 3). The rates of the microbial destruction of petroleums vary over three orders of magnitude. But even crude estimates of environmental persistence are of value in formulating the initial simulation models of a pollutant's disposition in various reservoirs—levels that can be checked in subsequent surveillance programmes.

Bioaccumulation is more amenable to laboratory studies than is degradation. For example, the concentration of ruthenium by some sea-weeds was first noted in the laboratory. Field observations are necessary, however, for those organisms which for one reason or another are not maintainable in laboratory aquaria. The enrichment of plutonium in giant algae (see

Chapter 4) was discovered through systematic studies of the concentration of the transuranic in marine organisms.

Toxicological studies can be based upon impacts on organisms or upon human health through the ingestion of sea foods. The latter guide has been effectively employed in regulating environmental levels of artificial radionuclides while the former has been used in restricting the utilization of DDT as a biocide. Where possible exposure levels are predicted to be close to toxic levels, regulatory action regarding the pollutant should be considered by those responsible for the management of the environment.

One of the present exasperating problems in pollutant studies is the retrieval of substantial information. Relevant data often exist but are elusive; hidden in file drawers, in company or government publications of limited distribution, or even in the open literature (poorly indexed), their recovery takes the form of a detective story. Solutions to this information problem are not yet at hand.

The geography of pollution

As previously mentioned, an approximate but useful measure of the potential pollution of a country is given by the ratio of its gross national product (GNP) to its area. This ratio spans a range of over four orders of magnitude (Tables 42 and 43) and can be correlated to the anthropogenic leaks of materials to the environment. The ranking of countries on the basis of the GNP: area ratio identifies problem areas. Japan and the United States are near the top of the list while the newly developing African and South American countries appear lower down. In general, northern hemisphere countries display higher ratios than southern hemisphere countries.

Some pinpointing of potential pollution

TABLE 42. Ratios of GNP to national area ranked in decreasing order for the countries in the northern hemisphere

Country	GNP (10^9/yr)	Population (millions)	Area (10^3 miles^2)	GNP/area (10^6/yr/miles^2)
Hong Kong	2.6	4.05	0.4	6.5
Singapore	1.3	2.11	0.224	5.8
Malta	0.275	0.33	0.122	2.25
Netherlands	1.78	13.2	13.967	1.78
Belgium	17	9.73	11.8	1.44
Japan	200	104.66	142.74	1.40
Federal Republic of Germany	130	61.68	95.96	1.35
Switzerland	17	6.23	15.94	1.06
Puerto Rico	3.5	2.84	3.44	1.017
United Kingdom	96	55.35	94.50	1.015
Luxembourg	1	0.34	0.999	1.001
Barbados	0.139	0.24	0.166	0.839
Italy	72	54.08	116.3	0.619
Denmark	9.8	4.92	16.61	0.590
German Democratic Republic	21	17.25	41.65	0.504
France	100	51.28	212.21	0.471
Trinidad and Tobago	0.88	1.07	1.98	0.444
Kuwait	3	0.83	7.68	0.39
Lebanon	1.4	2.87	4.05	0.345
Czechoslovakia	17	14.5	49.37	0.344
Taiwan	4.3	14.35	13.81	0.311
Algeria	4.3	14.77	919.591	0.291
Romania	26.6	20.47	91.7	0.29
Hungary	9.3	10.36	35.9	0.259
United States	930	207.01	3,615.12	0.257
Poland	28	32.75	120.35	0.232
Jamaica	0.92	1.86	4.41	0.208
Sweden	25	8.11	173.67	0.1439
Portugal	5.1	9.63	35.51	0.1436
Bulgaria	5.9	8.49	42.8	0.1378
Spain	25	34.13	194.88	0.128
El Salvador	1	3.53	8.06	0.124
Greece	6.2	8.96	50.94	0.121
Yugoslavia	11	20.55	98.77	0.111
Ireland	2.6	2.97	27.14	0.0957
Sri Lanka	2.1	12.51	25.33	0.0829
Dominican Republic	1.5	4.01	18.7	0.0802
Norway	9	3.88	125.181	0.0718
Democratic People's Republic of Korea	3.2	14.28	46.81	0.0683
Cuba	2.85	8.55	44.23	0.0643
Philippines	7.2	38.49	115.74	0.0622
Finland	7.9	4.68	139.12	0.0567
Republic of South Viet-Nam	3	18.33	65.73	0.04564
Guatemala	1.9	5.35	42.04	0.04519
Pakistan	14	58.7	342.86	0.04083
Costa Rica	0.74	1.69	19.7	0.03756

TABLE 42. Ratios of GNP to national area ranked in decreasing order for the countries in the northern hemisphere (*continued*)

Country	GNP (10^9/yr)	Population (millions)	Area (10^3 miles^2)	GNP/area (10^6/yr/miles^2)
Qatar	0.15	0.08	4	0.0375
Venezuela	0.4	9.4	252.14	0.03728
India	45	550.37	1,261.8	0.0356
Democratic Republic of Viet-Nam	2.25	21.6	63.36	0.0355
Turkey	10.5	36.16	301.38	0.0348
Haiti	0.345	4.97	10.71	0.0322
Mexico	24	50.83	761.53	0.0315
U.S.S.R.	240	241.75	8,599.34	0.0279
Panama	0.83	1.415	29.21	0.0284
Ghana	2.5	8.55	92.1	0.0271
Thailand	4.8	34.15	198.27	0.0242
Malaysia	3	10.80	127.67	0.0234
Canada	75	21.41	3,851.81	0.01947
China	70	787.2	3,691.5	0.0189
Nepal	1	10.845	54.6	0.0183
Morocco	3	15.31	171.4	0.0175
Nigeria	6	56.51	356.669	0.0168
Syria	1.2	6.45	71.498	0.0167
Nicaragua	0.933	1.98	57.15	0.0163
Honduras	0.674	2.58	43.27	0.0155
Sierra Leone	0.427	2.6	27.93	0.0152
Jordan	0.567	2.38	37.73	0.0150
Burma	3.9	27.6	261.8	0.0148
Colombia*	6.5	21.77	439.7	0.0147
Egypt	5.5	34.13	386.2	0.0142
Republic of Korea	0.545	32.43	38.45	0.0141
Senegal	0.79	3.78	76.12	0.01378
Iraq	2.3	9.75	167.6	0.01372
Cambodia	0.91	6.7	69.9	0.0130
Iran	8.2	29.78	636.4	0.0128
Togo	0.27	1.91	21.85	0.0123
Ivory Coast	1.5	4.42	123.5	0.0121
Uganda*	1.1	10.13	91.13	0.01207
Iceland	0.479	0.21	39.7	0.01206
Liberia	0.46	1.17	43	0.01106
Gambia	0.032	0.37	4	0.008
Belize	0.065	0.13	8.9	0.00737
Central African Republic	0.2	1.64	240.5	0.0068
Kenya*	1.5	11.69	224.96	0.0066
Cameroon	0.9	5.84	183.6	0.0049
Afghanistan	1.2	17.48	251	0.0047
Benin	0.2	2.76	44.7	0.0044
Ethiopia	1.8	25.05	471.78	0.0038

* These countries lie exactly on the Equator, but since more than 50 per cent of their areas lie north of the Equator they are assumed to be countries belonging to the northern hemisphere.

TABLE 42. Ratios of GNP to national area ranked in decreasing order for the countries in the northern hemisphere (*continued*)

Country	GNP (10^9/yr)	Population (millions)	Area (10^3 miles^2)	GNP/area (10^6/yr/miles^2)
Surinam	0.218	0.41	63.25	0.0034
Guyana	0.266	0.763	83	0.0032
Guinea	0.28	4.01	94.93	0.0029
Equatorial Guinea	0.04	0.29	17.53	0.0028
Laos	0.202	3.03	91.42	0.0022
Democratic Yemen	0.016	1.47	112.08	0.00142
Yemen	0.127	5.9	75.29	0.00113
French Guiana	0.032	0.051	35.14	0.00091
Saudi Arabia	0.697	7.2	873.97	0.000797
Mongolia	0.52	1.29	604.25	0.00086
Niger	0.321	4.13	459.07	0.000699
Mali	0.245	5.14	464.88	0.000527
Chad	0.25	3.8	490.7	0.000509
Bangladesh	0.005	75	55.1	0.00009
Upper Volta	0.003	5.49	105.84	0.000028
Tunisia	0.001	5.14	63.38	0.000017
Libyan Arab Republic	0.002	2.01	679.36	0.000002

sources is indicated from this approach. The island emerges as a fragile land type, especially susceptible to possible pollution. For the northern hemisphere, the three countries with the highest value of the GNP/area ratio, i.e. those subject to more serious pollution problems than those with lower values of the ratio, are all islands—Hong Kong, Singapore and Malta (Table 42). Within the top ten countries, ranked in order of decreasing values of the ratio, six are islands, Puerto Rico, Japan and United Kingdom joining the above-mentioned trio. Of the ten southern hemisphere countries with the highest values of the ratio, five are islands, Nauru, Mauritius, New Zealand, Fiji and Indonesia (Table 43).

The implications of this observation are clear. The relation of the coastline area of an island country to its total area is high compared to countries situated on large continents. The coastal zone is a primary resource of most countries. Demands upon its use take a variety of forms—recreational facilities for the inhabitants as well as for tourists; manufacturing industries, which can take advantage of harbour facilities for importing or exporting raw materials and finished products and which can utilize the sea waters as coolant waters or receptacles for wastes; and commercial sea-food industries, which require space for the harvesting or cultivation of sea foods. The resolution of conflicts in such uses is a primary responsibility of coastal zone managers.

Developing island nations can well look at technologically advanced island nations for the patterns of marine pollution that can potentially develop. Perhaps Japan serves as the 'test tube' for studies of pollution of the sea. Here, a densely inhabited country allowed conflicts to develop in the use of the coastal zone. Industrial effluents discharge into areas where sea foods are harvested. Wastes from manufacturing are indiscriminately dumped into nearshore

Table 43. Ratios of GNP to national area ranked in decreasing order for the countries in the southern hemisphere

Country	GNP (10^9/yr)	Population (millions)	Area (10^3 miles^2)	GNP/area (10^6/yr/miles^2)
Nauru	0.025	0.007	0.008	3.125
Mauritius	0.148	0.82	0.72	0.205
Rhodesia	1.2	5.27	150.3	0.117
New Zealand	5.5	2.85	103.8	0.052
Fiji Islands	0.269	0.53	7.1	0.037
South Africa	15.3	21.28	471.4	0.032
Uruguay	1.6	2.29	72.2	0.022
Rwanda	0.21	3.59	10.2	0.020
Indonesia*	12	124.89	735.3	0.016
Chile	4.6	8.83	292.3	0.015
Argentina	14.8	23.4	1,072.07	0.0138
Burundi	0.145	3.62	10.7	0.0135
Swaziland	0.085	0.42	6.7	0.0126
Ecuador*	1.87	6.3	175.9	0.0106
Australia	26.4	12.7	2,967.91	0.0088
Lesotho	0.094	1.04	11.7	0.008
Malawi	0.272	4.55	36.5	0.0074
Brazil*	24	92.23	3,286.5	0.0073
Zambia	1.6	4.3	290.6	0.0055
Paraguay	0.587	2.38	157	0.0037
Madagascar	0.78	6.75	226.7	0.0034
Tanzania	1.2	13.63	362.8	0.0033
Angola	1.2	5.43	481.35	0.00249
Bolivia	0.951	5.06	424.2	0.00224
Zaire*	1.5	22.48	905.1	0.00165
South-West Africa	0.5	0.65	318.3	0.00157
Mozambique	1.2	7.38	785	0.00152
Congo*	0.2	0.96	134.8	0.00148
Peru	0.49	14.01	496.2	0.00098
Botswana	0.08	0.7	222	0.00039
Gabon*	0.001	0.5	102.3	0.0000097

* These countries lie exactly on the Equator, but since more than 50 per cent of their areas lie south of the Equator they are assumed to be countries belonging to the southern hemisphere.

waters. As a consequence, disasters have occurred, and threats to human health have evolved. It is interesting to note that many prominent environmental diseases have Japanese names: Minamata Bay Disease (methyl mercury poisoning); Yusho (PCB poisoning); and Itai-itai (cadmium poisoning).

The relative degree of pollution of marine areas

Some measure of the relative degree of pollution of marine areas can serve as a guide for national and international action, both regulatory and scientific. Widely varying social patterns of coastal populations, coupled with the unique character of each coastal area, makes general approaches rather difficult to formulate. There is a great need for predictive abilities with respect to the vulnerable estuaries and semi-enclosed nearshore areas. Even approximate measures of pollution can be useful in defining the need for protective measures.

A measure of the relative degree of pollution in a semi-closed marine area like the North Sea, Chesapeake Bay or the Inland Sea of Japan is sought. Let us call this measure P. To a first approximation, P will be a function of the mixing or flushing time of the water body with the open ocean, t, and of the flux of anthropogenic materials entering the basin, F. Then, we can write: $P=P(t, F)$, assuming that the biological, chemical and physical processes that act upon the collective of pollutants are the same for all areas under consideration.

For most basins, t is known to within a factor of approximately 2. The determination of the total anthropogenic flux will be more difficult to ascertain. Entry to the basin will be divided among the various transport paths, atmosphere, rivers and outfalls. Pollutant fluxes along these paths are not known for any of the major coastal areas of the world. Hence, recourse must be had to an indirect estimate.

The *per capita* GNP of the population whose wastes are in part discharged to the waters is a useful indicator of the weighting factor to be applied to the number of people in the system. Thus, for any country (i) bordering the area, its flux of material, F_i, to the waters is given by: $F_i=F_i\,(\text{GNP}_i \times N_i)$ where GNP_i is its *per capita* gross national product and N_i is its population. For all of the countries bordering the zone,

$$F=k\sum_i [(\text{GNP})_i \times N_i],$$

where k is a proportional factor with units of cubic kilometres per dollar days.

If we also assume that the mixing or

flushing processes are the only processes that remove the pollutants from the zone, then the total amount of pollutant material, M, in the waters is related to the flushing time and flux by: $M=t \times F$.

$$P=M/V=t \times k \times \sum_i [(\text{GNP})_i \times N_i]/V.$$

Rankings of the pollution for several semi-closed marine areas will be given to illustrate the use of this equation. Our measure of pollution will be P/k.

Puget Sound, Washington (United States) (data supplied by Professor Roy Carpenter, University of Washington, Seattle, Washington): volume, 122 km³ (V); flushing time, 152 days (t); N_i, 2,000,000 (United States only); GNP_i, \$5,000; P/k, 1.25×10^{10}.

Chesapeake Bay (United States) (data supplied by Dr Harry H. Carter, Johns Hopkins University, Baltimore, Maryland): volume, 74 km³ (V); flushing time, 209 days (t); N_i, 7,000,000 (United States only); GNP_i, \$5,000; P/k, 10×10^{10}.

Thames Estuary, British Isles (data supplied by P. C. Wood, Ministry of Agriculture, Fisheries and Food, Burnham-on-Crouch, United Kingdom): volume, 19 km³ (including inner and outer estuary) (V); flushing time, 10 days (t); N_i, 15,000,000; GNP_i, \$1,700; P/k, 1.3×10^{10}.

North Sea (data supplied by P. C. Wood, Ministry of Agriculture, Fisheries and Food, Burnham-on-Crouch, United Kingdom): volume, 54,000 km³ (V); flushing time, 1.5 years (t);

Country	N_i	GNP_i	$N_i \times \text{GNP}_i$
Norway	2.3×10^6	2,300	5.3×10^9
Sweden	1.1×10^6	3,100	3.4×10^9
Denmark	0.6×10^6	1,950	1.2×10^9
Federal Republic of Germany	2.9×10^6	2,170	6.3×10^9
Netherlands	2.2×10^6	1,890	4.2×10^9

Country	N_i	GNP_i	$N_i \times GNP_i$
Belgium	0.25×10^6	1,750	0.4×10^9
France	0.6×10^6	1,930	1.2×10^9
United Kingdom	21×10^6	1,700	35.6×10^9
Total	31×10^6		57.6×10^9

P/k, 5.8×10^8.

Gulf of Mexico (data supplied by Professor William Sackett, Texas A and M University, College Station, Texas): volume, 2.29×10^6 km^3 (for total Gulf) (V); flushing time, 100 years (for waters deeper than 1,800 m. 2,000 m is mean depth) (t); N_i, 30,000,000 (United States only); GNP_i, \$5,000; P/k, 1.5×10^9.

A second calculation can be made using only the shelf waters: volume of shelf waters: 3.1×10^4 km^3 (V); flushing time, 20 years (t); P/k, 2.3×10^{10}.

The values for the Thames Estuary and for Chesapeake Bay are probably high in comparison to the others inasmuch as these water bodies receive sewage effluents which in general have been subjected to secondary or tertiary treatment.

The model suggests, as an example of the type of information to be gleaned, that the North Sea is less vulnerable to an alteration in chemical composition than the other waters considered, primarily as a result of its short flushing time and of its large volume. This conclusion is in agreement with recent reports (ICES, 1974, and Goldberg, 1973) that the North Sea's open waters do not contain significant levels of pollution from terrestrial discharges and is in conflict with arguments that the North Sea is highly polluted (Weichart, 1973).

An example

Prediction of future exposure levels of a potential marine pollutant can often be made with very little data. Such is the case for perchloroethylene (per) whose global production, environmental persistence and toxicity suggest that a continuing knowledge of its concentrations in our surroundings is called for. The following development is taken from NAS, 1975.

The United States production of per in 1972 was 333,000 tons. A global production of a megaton per year can be extrapolated from this figure. Production increased at an annual rate of 9 per cent in the United States between 1962 and 1972. A global production increase of 5–7 per cent annually over the next few years is reasonable. The primary use of per is in drycleaning, which consumes about two-thirds of the production. Its other principal industrial uses are in vapour degreasing and as a chemical intermediate.

Per has well recognized anesthetic properties to both humans and animals, and at high concentrations its ingestion can cause liver and kidney damage. Like most toxic chemicals, there is no information on long-term, low-level biological effects.

There are conflicting reports about the photochemical stability of per in the atmosphere. Altschuller *et al.* (1971) found it to be virtually unreactive, a result contrary to those of the Dow Chemical Company reported in NAS, 1975. The latter investigation found the photochemical oxidation of per in simulated sea-level experiments was rapid with half-lives of the order of hours.

Loss from natural waters may be due to evaporation, photochemical decomposition, biological degradation or chemical hydrolysis. The Dow Chemical results indicated no biological oxygen demand for per within a period of twenty days. However, in aerated waters, Dow Chemical found that after exposure in the dark for six months 63 per cent of per at an initial concentration of 1 p.p.m. remained, while 52 per cent remained when exposed to light under the same conditions.

The environmental measurements cited in Chapter 3 may now be coupled to the production data to construct a provisional flow diagram for per. The mean atmospheric concentration is taken as 5×10^{-9} g/m³. Since essentially all of the release is to the northern hemisphere, its atmosphere then contains 10^4 tons, assuming a uniform distribution. The global flux, essentially restricted to the northern hemisphere, is 9×10^5 tons per year, assuming that all uses of per are dispersive except where it is used as a chemical intermediate, which takes up about 10 per cent of the total production. The residence time in the atmosphere, defined as the steady-state concentration divided by the rate of throughput, is then of the order of four days. The removal of per from the atmosphere with such a time constant is most probably due to photochemical decomposition.

The mean concentration of per in surface waters of the North Atlantic is about 0.5 ng/l (about 3×10^{-9} the value of its solubility). On the basis of its atmospheric concentration of 5 ng/m³ (about 3×10^{-11} times the saturation concentration at 20° C), NAS (1975) suggests a net transfer to the atmosphere across the ocean surface. Clearly, these data, based upon analyses primarily in near-coastal waters, may not be representative of the entire northern hemisphere oceans. NAS (1975) indicated that perhaps 3 or 4 per cent of the total annual production enters the ocean surface.

This would be then 3×10^4 tons per year. A mean concentration of 0.5 ng/l in the upper 40 metres of the northern hemisphere oceans gives a burden there of about 3,000 tons and a residence time of about one month.

Thus, the preliminary flow diagram would take the form shown below.

Clearly, some of the per introduced into the surface layers of the ocean is transferred to the deeper layers. Also, some of the per in the surface layers is decomposed either inorganically or biologically. Both processes would have a tendency to reduce the surface water residence time of one month. As it now stands, evaporation appears to be the primary way of removing per from surface waters.

The model indicates that per is primarily derived from direct discharge into the oceans—river runoff, sewer discharge or dumping. Accordingly, strong concentration gradients decreasing seaward from the injection sites in the coastal zone are expected.

The model illustrates the types of initial inductions that can be derived from limited data. They clearly provide guidance for an initial surveillance programme to establish or deny the validity of the model. The model also provides a method for the initial prediction of future exposure levels. Assuming the present levels are nearly at steady state, since the atmospheric and oceanic residence times are of the order of days and months, future levels can be extrapolated linearly from present levels with estimated changes in production with time.

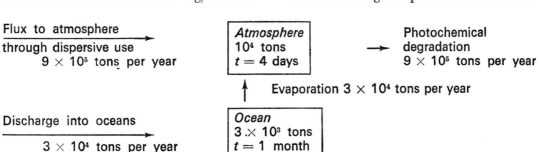

Bibliography

ALTSCHULLER, A. P.; LONNEMAN, W. A.; SUTTERFIELD, F. D.; KOPCZYNSKI, S. L. 1971. Hydrocarbon composition of the atmosphere of the Los Angeles Basin—1967. *Envir. Sci. Technol.*, vol. 5, p. 1009–16.

GOLDBERG, E. D. 1973. *North sea science.* Cambridge, Mass., MIT Press. 500 p.

ICES. 1974. Report of Working Group for the International Study of the Pollution of the North Sea and its Effects on Living Resources and their Exploitation. *Cooperative Research Report No. 39.* Charlottenlund, Denmark, International Council for the Exploration of the Sea. 191 p.

NAS. 1975. *Assessing potential ocean pollutants.* Washington, D.C., National Academy of Science. 438 p.

WEICHART, G. 1973. Pollution of the North Sea. *Ambio*, vol. 2, p. 99–106.

9. Monitoring strategies

The goal of monitoring programmes is to provide governmental agencies with a scientific basis to regulate the release of materials that may have a deleterious effect upon marine resources. Most, but not all, monitoring efforts and regulatory activities have been aimed at the maintenance of public health, which might be jeopardized by the consumption of polluted sea foods or by direct exposure to a toxic substance. While the regulation of radio-active material discharge has been based upon possible returns to man of ionizing radiation, the banning of DDT by some countries was due to its toxic effects, where used as a pesticide, upon non-target organisms, not including man.

The results of monitoring studies can be used to assess the effectiveness of existing regulatory activities. Since there are economic limitations to the number of samples that can be examined, whether based upon type, location or frequency of analysis, selecting and obtaining pertinent information has very high priority in the design of a surveillance programme. The 'critical pathways' approach, developed for the monitoring of radio-activity in marine systems, has been most successful in minimizing sample numbers. It is based upon the argument that of the many complex pathways by which a pollutant released to the environment can return to man, there are one or two pathways that are more important than all others. It is the identification of these critical pathways that can lead to a rational monitoring programme.

However, inadequacies in the critical pathways approach may be revealed in time if its primary assumption proves erroneous. Hence, a more systematic surveillance programme may be required to reveal the dispersion of a pollutant about the environment. The 'mass-balance approach', discussed previously, provides a broader understanding of pollutant distribution. For this approach, a monitoring programme is required to assess the mass-balance calculations and to provide inputs for their formulation. As a consequence, the primary sites of accumulation of pollutants can be identified, as well as the fluxes into, and within, the marine environment.

The critical pathways approach

The basis of the 'critical pathways approach' rests in the development of acceptable exposure criteria for the pollutant under consideration (Preston and Wood, 1971). However, established permissible levels are usually not without risk. An 'acceptable dose' on the basis of the average exposure of individuals usually entails a certain risk to a segment of the population whose life styles subject them to a far greater exposure than that of the normal individual.[1] Let us consider the following hypothetical example. A pollutant is found to concentrate in commercial fish. The average consumption of fish for the entire population is 50 grammes per day per person; yet some fishermen, say 0.001 per cent of the entire society, eat more than ten times this amount daily (300 grammes constitutes an average weight of fish for an adult meal). Thus, the acceptable exposure level involving the consumption of the pollutant may be based upon providing minimum risk to 99.999 per cent of the population or to the entire population. In some cases it may be more reasonable to seek a change in the eating habits of the fishermen than to maintain an unrealistic 'acceptable dose', which is economically difficult to achieve. Such considerations are of importance in formulating policy to minimize the risk to an individual of experiencing an ill effect during his lifetime through

exposure to a pollutant: the somatic risk.

The risk entailed by possible damage to the gene pool of a population, i.e. chromosomal alteration, may be transferred to a future generation. This genetic risk is based primarily upon the average *per capita* dose and the total number of people exposed rather than the amount of a dose to an individual. In general, genetic dose limits have lower values than their somatic counterparts. For example, in radiation doses, both internally and externally, the individual or somatic risk value is 0.5 rem per year per whole body, while for populations a value of 5 rem per person per 30 years is recommended. Both of these guides have been formulated by the International Commission on Radiological Protection.

In most cases a pollutant will take a critical path back to man through sea foods. Eating habits of the concerned populations must be used to identify the highest consumers of the food containing the toxic substances, the 'critical group'.

A priori calculations of expected exposure levels by man can lead to a rational monitoring programme using the critical pathways approach. The steps in such a procedure have been outlined by Preston and Wood (1971) in the following way:

1. Estimation of the pollutant concentration in the sea water system as a function of the rate of introduction of pollutant.
2. Estimation of the concentration factor of the pollutant in the critical material, be it a living organism or sediment.
3. Coupling 1 and 2 leads to an estimation of the concentration of the pollutant in the critical material.

---

1. The concept of acceptable dose also entails consideration of the protection of perhaps more vulnerable elements in the population such as babies, pregnant women and elderly people. Such individuals may be more sensitive to the intake of a given pollutant than other members of society.

This preoperational calculation may be compared with:

4. The acceptable daily intake of, or exposure to, the pollutant by man, which is determined by
5. An acceptable concentration in the critical material.

A comparison of the value obtained in 5 with that in 3, leads to a permissible rate of introduction of the pollutant under 1. The validity of the preoperational assessment can then be checked by the monitoring programme.

An example of the actual use of the 'critical pathways' approach involves the movements of the radio-isotope ruthenium-106, a fission product which accumulates in fuel elements of nuclear reactors and is discharged to coastal waters from the United Kingdom reprocessing plant at Windscale. The nuclide is concentrated by the sea-weed *Porphyra* which is incorporated in a foodstuff called laverbread consumed by some people in southern Wales. A detailed description of this problem is provided by Preston and Mitchell (1973) from which the following summary has been obtained.

The number of laverbread consumers is estimated to be about 26,000, all of whom live in southern Wales near the sites of production of this food. The eating patterns of the population were surveyed in 1967 and it was found that a small group ate the sea-weed product at an extremely high rate. This subset of the population forms the so-called 'critical group' and its mean consumption rate was 160 grammes of laverbread per person per day.

The internal exposures to ionizing radiation from the consumption of laverbread are not only due to the ruthenium-106 but to other radio-active species in the discharges. In dosages to the gastrointestinal tract the ruthenium-106 constitutes about 90 per cent of the radio-active species received with the remainder being contrib-

uted equally by cerium-144 and by the pair zirconium-95 and niobium-95. Smaller doses are received by the bone primarily resulting from the uptake of strontium-90, plutonium-238, -239, -240 and americium-241.

The mode of production of the laverbread leads to some uncertainties in the estimation of exposure levels. Usually, the laverbread is made in part from *Porphyra* harvested near the Windscale discharge. The potentially polluted materials are often diluted with inactive weeds from other areas of the British Isles. Occasionally, pure 'Windscale' laverbread is marketed.

Monitoring of the polluted sea-weed and of the actual product has been found to be essential, primarily because of the doubt associated with dilution factors with unpolluted *Porphyra*. Monitoring of the levels in the sea-weed have extended in distances up to 40 kilometres from the discharge point. The exposure levels in the lower large intestine are shown in Figure 24 where dose rates have been calculated on the basis of the ruthenium-106 concentrations in the contaminated sea-weeds and in the product itself. The dilution factors are quite substantial and the estimated critical organ dose is usually under 1 per cent of that recommended by the ICRP for the critical population eating a mean amount of 160 grammes per day.

Only one external pathway of radioactive nuclides back to man has so far been identified (Preston and Mitchell, 1973). This concerns the removal of gamma-emitting radionuclides to sediments. The exposed population includes fishermen who spend time on mud banks. Exposure estimates have been based upon a single individual who fishes about 300 hours per year. For the years 1969–71, his estimated exposure ranged between 7 and 18 per cent of the ICRP recommended limits.

The 'critical pathways' approach does

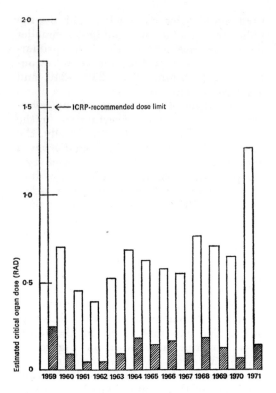

FIG. 24. Radiation exposures of the lower large intestine of consumers of laverbread from the consumption of 160 grammes per day. Unshaded blocks: dose rate assuming consumption of laverbread made entirely from polluted Windscale sea-weed; shaded blocks: dose rates based upon consumption of laverbread made from both polluted and unpolluted sea-weeds. Dilution data for 1959–65 estimated from shipment records; data for 1966–71 are from actual laverbread analyses (Preston and Mitchell, 1973).

not generally require sea-water analyses, which can be both difficult to make and expensive.

Mass-balance approach

In order to describe present dispersions of a pollutant in the world oceans and to predict future ones, models must be developed and tested to relate quantitatively the source functions, reservoir concentrations, fluxes between reservoirs and sinks (see Chapter 2). These models that balance inputs and outputs are known as 'mass-balance models' and can be developed for either the steady-state situation or for the transient situation. They can be placed in the following general form $\Delta C = C_a + C_b + C_c - C_d - C_e$; where ΔC is the annual change in the pollutant content of the open ocean; C_a is the flux from the coastal ocean to the open ocean in grammes per year; C_b is the flux from the atmosphere to the open ocean in grammes per year; C_c is the flux from ships to the open oceans in grammes per year; C_d is the chemical (or radio-active) disintegration in the open ocean in grammes per year; C_e is the flux to the sediments from the open ocean in grammes per year. For a given pollutant, one or more of the terms on the right-hand side of the equation may be trivial. For example, in the case of DDT and its residues, C_c and C_a are small compared to C_b and may be neglected.

These models and their mass-balance calculations are an integral part of a global monitoring programme. The assessment of their validity depends upon the availability of adequate field data and production and use data. For the field programme, the following characteristics derive from scientific and economic arguments:

The smallest number of samples consistent with ensuring the statistical validity of the model should be sought for the pollutant under consideration.

Since the risk of contamination may be great for many oceanic and atmospheric pollutants of low concentration, sample collections should only be made by qualified personnel aware of these difficulties.

A small number of laboratories should be involved in the analytical activities. Expensive, sophisticated equipment is often necessary. The number of competent analysts is low.

Since there is a greater use of materials and production of energy in the northern hemisphere compared to its southern counterpart, sampling emphasis should be on the former area.

An approach to a global monitoring programme that might satisfy the needs of a mass-balance approach without unreasonable financial and technological strains has been proposed by Goldberg *et al.* (1971). A 1,000-sample exercise to provide baseline data for a monitoring programme was formulated. It was suggested that the samples could be gathered and analysed over a year. The distribution of samples is illustrative of how global monitoring systems might be devised.

Atmosphere (202 samples). Air from the principal wind systems (tradewinds, mid-latitudinal westerlies and polar easterlies) would be sampled at five elevations including the surface (Table 44). Continental and marine stations are envisaged.

Northern rather than southern hemisphere sites would be emphasized.

Ocean current system (140 samples). Three types of vertical profiles were suggested. The first is simply a sample of surface film. The second includes the surface film, the mixed layer and the pycnocline. The third adds to the samples of the second, a set from 1,000, 2,000 and 3,000 metres and near bottom. The fifty surface film samples should be taken from the world ocean with emphasis on the major zones of precipitation and evaporation, and along the major shipping routes (especially for petroleum-derived hydrocarbons). For the shallow arrays, sampling was proposed for the major eastern and western boundary currents (about sixteen positions) and the eastern and western edges of the Atlantic, Pacific and Indian oceans at the Equator. With three samples per array and twenty-two arrays, a total of sixty-six samples is suggested. For the deep arrays, the sampling sites would include the centres of

TABLE 44. Sampling network for atmospheric pollutants

Location	Number of samples				Frequency per year	Total
	Surface	500 m	1,500 m	5 km		
North America	15	7	4	2	2	56
Europe and Asia	20	10	5	3	2	76
Africa	3	0	3	3	1	9
South America	3	0	3	3	1	9
North Atlantic	5	0	3	3	1	11
South Atlantic	3	0	3	3	1	9
North Pacific	7	0	4	4	1	15
South Pacific	3	0	0	3	1	6
Australia	1	0	1	1	1	3
Arctic	1	0	0	1	1	2
Antarctic	1	0	0	1	1	2
Subtropical high-pressure cells	2	0	0	2	1	4
TOTAL						202

Reproduced from E. D. Goldberg *et al.*, 'Proposed Baseline Sampling Program', in *Chlorinated Hydrocarbons in the Marine Environment*, p. 23–6, Washington, D.C., National Academy of Sciences, 1971.

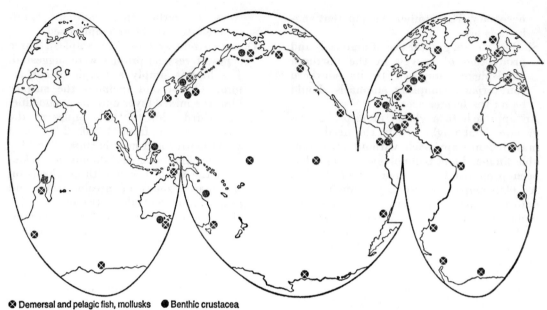

⊗ Demersal and pelagic fish, mollusks   ● Benthic crustacea

FIG. 25. Fish, mollusk and crustacea sampling areas.
(From E. D. Goldberg *et al.*, 1971.)

the major gyres and the major semi-enclosed seas (Arctic, Gulf of Mexico, Mediterranean, Okhotsk, Bering and Sea of Japan). A central equatorial array for each ocean is also proposed. A total of twenty deep arrays with seven samples each gives a total of 140 samples.
Organisms (320 samples). Fish, 124; molluscs, 42; crustacea, 12; and plankton, 142. Areas of high primary productivity and of major fisheries were emphasized inasmuch as they may provide food pathways for the pollutants leading to man. The sampling sites shown in Figure 25 take into account the importance of the major fishing areas of the North-East Atlantic, West central Pacific and South-East Pacific which accounted for 61 per cent of the 1967 marine fishery (see Table 45). It is interesting to note that more than 65 per cent of the Atlantic catch was taken above 30º N., the latitudinal zone of the major sources of pollution.

Rivers and continental-shelf areas (126 samples). Rivers were selected on the basis of the following criteria: large size; extensive industrial and/or agricultural activity; representation of different climatic zones; and representation of different stages of industrial development. Two sampling periods, high-flow and low-flow stages, were proposed with collections of both dissolved and particulate phases. The surface sample would collect primarily the dissolved phases while the bottom sample would collect water and sediment. Nineteen river systems were designated from which four samples per year would be taken (seventy-six total river samples): North America—Hudson, Mississippi, St Lawrence and Columbia; South America—Orinoco, Amazon and Plata; Africa—Nile, Congo and Niger; Europe—Rhine, Danube, Po and Thames; Asia—Ganges-Brahmaputra, Yellow, Amur, Ob and Indus. Fifty sedi-

TABLE 45. Fish landings for 1967

Area	Millions of tons	Total catch (%)	Area	Millions of tons	Total catch (%)
North-West Atlantic	4	7.6	North Pacific	6.4	12.2
North-East Atlantic	10.2	19.5	West Central Pacific	10.5	20.0
West Central Atlantic	1.3	2.5	East Central Pacific	0.7	1.3
East Central Atlantic	1.6	3.1	South-East Pacific	11.2	21.5
South-West Atlantic	1.3	2.5	South-West Pacific	0.4	0.8
South-East Atlantic	2.5	4.8	West Indian Ocean	1.3	2.5
			East Indian Ocean	0.8	1.5

ment samples from continental shelves adjacent to estuaries or river deltas were proposed to understand the flow of particulates with their associated pollutant materials into the open-ocean areas. Floods usually are the triggering actions for such movements. Eleven areas were chosen from which an estimated fifty samples would be taken annually: Yellow Sea, Mississippi Delta, Nile Delta, Baltic Sea, Sea of Japan, Ganges-Brahmaputra Delta, Tokyo Bay, Osaka Bay, North Sea, Black Sea and Rhone Delta.

Rain (48 samples). Precipitation washout measurements are proposed utilizing rain samples collected from each of the trade-wind and jet-wind systems in the northern hemisphere. Monthly composite samples integrated over a year would be useful for comparing with the record in the glaciers (see following discussion).

Glaciers (78 samples). The atmospheric injections into the open ocean may be recorded in permanent snowfields (glaciers) for fallout from all of the major wind systems (see Chapter 2): Greenland, northern hemisphere polar easterlies; Yukon territory, northern hemisphere jets; Mount Orizaba or Mount Popocateptl, Mexico, northern hemisphere trades; Andean glaciers, southern hemisphere trades; Tasman glaciers, New Zealand, southern hemisphere jets; and Antarctica, southern hemisphere polar easterlies.

Deep-sea sediments (24 samples). Sampling of deep-sea sediments, especially those containing fossil remains of organisms, provides a measure of the removal of materials from surface waters to the deposits by biological agencies. A suite of four samples each of siliceous and calcareous ooze from the Pacific, Indian and Atlantic can provide a reasonable set of materials to investigate biological removal mechanisms.

Such a sampling programme clearly would be modified as results were obtained, with the aim of retrieving the smallest number of samples consistent with fulfilling the needs and the aims of the model under assessment. Because of the wide dispersal of sampling sites, hot spots, localized areas of extremely high concentrations of a pollutant, may be missed. Such areas would probably be close to the sources of the pollutant, in coastal regions. Still, a 1,000-sample, mass-balance modelling will be expensive to undertake. A less costly monitoring programme to establish pollutant trends (described in Chapter 10) can be a forerunner of this type of activity to establish the need for a more detailed analysis of pollutant dispersal.

Biological indicators of pollution

A single species or a group of marine organisms may be open to physiological or behavioural disturbances through exposure to a pollutant. In principle, such an effect can be detected by the long-term changes in community structure. In practice, however, monitoring for such an effect is not yet possible, since such monitoring is fraught with difficulties.

First of all, the cause of a subtle population alteration is hard to assess. Even if there is background information collected over a sufficiently long period to establish the range of variation, it is at present not possible to determine whether any small variation is due to natural fluctuations of the population, to a climatic change, or to a pollutant.

Accurate estimates of biomass depend upon an adequate treatment of uneven, often patchy population distributions. At present, the distinction between temporal or spatial changes and random changes cannot be made for pelagic populations.

Some species have varying sensitivities to a pollutant during the different periods of their life cycles.

Synergistic or antagonistic effects of one pollutant upon another need to be considered; however, very little work has so far been carried out in this area.

Finally, the species most sensitive to a given pollutant are yet to be identified.

Although a large number of marine ecologists have intensively advocated the use of community structure assessment in monitoring programmes, there have so far been no demonstrable cases where pollutant effects have been established as the cause for modest changes in population. Mass mortalities, such as those due to the impact of petroleum spills on birds or of EDC tars on fish, have been recognizable. Perhaps other changes in living systems may turn out to be more rewarding than population studies, as for example chromosome aberrations, changes in behaviour patterns or physiology.

However, marine organisms are useful for estimating sea-water pollutant levels, for identifying potential health hazards through their ingestion, and for determining their role as a reservoir and transporting agency.

INDICATOR ORGANISMS

Certain organisms are especially susceptible to a given pollutant. For example, crustaceans react adversely to parts per billion levels of the heavier halogenated hydrocarbons. At higher levels of the food web, the reproductive success of birds feeding on marine fish suffered due to their ingestion of DDT residues. Although the 'most sensitive' species for a given pollutant may not be identified, species that are injured in one way or another by possible marine concentrations can be found through systematic laboratory testing programmes. The sensitivity may take the form of some behavioural response, growth inhibition, reproductive failure, or other types of dysfunction. The laboratory results can in principle be meaningfully extrapolated to the ocean environment.

BIOACCUMULATORS

Knowing the concentration factors for a given pollutant by a marine species and the concentration in indigenous organisms facilitates an estimation of the pollutant level in the environment from which the organisms were taken. In addition, predictions of future levels in the organisms can be made assuming a knowledge of future source functions and their associated environmental levels.

NON-INDIGENOUS ORGANISMS

The low populations of pelagic organisms makes significant assessments of pollutant impacts difficult to attain. An alternative tactic to the use of native organisms is the employment of non-indigenous organisms cultured on small islands, buoys, platforms or weather ships. Both bioaccumulators and indicator organisms may be transferred from their normal habitats to the pelagic zones of concern. Such organisms have not been identified hitherto; initially, mussels and barnacles might be considered for such a role. Through laboratory experiments it may be possible to determine any differences in the response to the pollutant by organisms in their native habitat and by those transferred to an alien environment.

COMMERCIALLY IMPORTANT SPECIES

Marine organisms of importance to man as food sources can carry potentially dangerous levels of pollutants. The 'critical pathways approach' seeks to monitor those species on the hopeful assumption that there will be but a few for each substance of concern. Also of importance is the stress upon the organism itself, its own body burden of pollutants. The surveillance of selected commercial species can thus provide a measure of their quality as food and a sense of their own well-being.

Berner *et al.* (1972) proposed a programme of biological monitoring in the coastal regions of the United States. Organisms were chosen on the basis of their importance to man, ease of sampling, ecological importance and biochemical, physiological and behavioural diversity. Their choice of species for monitoring provides a pattern for the development of more ambitious programmes. Their coastal zone selections include the following.

At the bottom of the food chain the molluscs *Mytilus* and *Rangia*, with wide geographic distributions, were chosen. These filter-feeding organisms can absorb pollutants directly from the water through their mucous sheets.

Herring, menhaden and anchovy are widely distributed plankton-feeding fish. Caught commercially, their high lipid contents may act as reservoirs for the halogenated hydrocarbons and petroleum.

The squid *Loligo* is a coastal zone cosmopolite, fished commercially and fed upon by commercially important fishes.

The commercially sought yellow-tail flounder and dover sole are bottom-feeding flat fish; the blue crab and Dungeness crab are also bottom feeders. Anadromous fish such as striped bass and silver salmon are top carnivores which may travel through highly polluted coastal areas in their paths from terrestrial waters to the open ocean. The silver salmon is important commercially while the striped bass is heavily fished by sports fishermen.

Laboratory housekeeping

The concentrations of the oceanic pollutants are often in the parts per billion or parts per trillion levels in waters, organisms, airs or sediments. The techniques of analysis are sophisticated, requiring talented workers and complex equipment. Contamination of these very low levels of pollutants can take place during sampling, storage or analysis. Precautions to minimize contamination must be taken at all stages of the work. The demanding nature of this type of assay has resulted in the development of only a dozen or so laboratories in the world capable of analysing such pollutants as the DDT residues, the PCBs or the transuranics in sea water. The levels of these compounds in organisms and sediments are usually higher and as a consequence, more laboratories can analyse such materials for oceanic pollutants.

Several tactics are available to aid laboratories in achieving accuracy in their results—the availability of standard reference materials, which have been accurately assayed for the pollutant in question, and interlaboratory analyses of commonly studied samples, such as sea water, or organisms such as herring and mussels. For nearly all pollutants the methods of assay, as well as those of collection and of preparation before analysis, vary from one laboratory to another. As a consequence of rapidly changing practices in the analytical chemistry of environmental samples, the

TABLE 46. Values reported for orchard leaves (NBS-SRM 1571)*

Element	Range of results**		Recommended interim value	NBS certified value
	NBS	Other		
Aluminium	341–428 (2)	99–420 (14)	400	
Antimony		2.7–3.5 (3)	2.9	
Arsenic	11.3–15.9 (2)	10–18.3 (7)		14 ± 2•
Barium		40–52 (5)	45	
Bismuth	0.10–0.14 (1)		0.11	
Boron	31–34 (1)	23.7–38 (13)		33 ± 3
Bromine	9.4–10.5 (1)	8.9–9.5 (2)	10	
Cadmium	0.09–0.12 (2)	<0.1–.45 (9)		0.11 ± 0.02
Calcium	2.052–2.125 % (2)	1.63–2.41 % (15)		2.09 ± 0.03 %
Chlorine	680–780 (1)	790 (1)	750	
Chromium	2.34–2.38 (1)	1.8–2.5 (2)	2.3	
Cobalt	0.18–0.23 (1)	0.1–0.18 (5)	0.2	
Copper	10.8–13.7 (3)	9.9–20 (25)		12 ± 1
Iron	282–319 (3)	151–367 (21)		300 ± 20
Lanthanum		1.2–2.1 (2)		
Lead	41.4–47.9 (3)	37–53 (12)		45 ± 3
Lithium	14 (1)		14	
Magnesium	0.6136–0.6370 % (1)	0.40–0.71 % (14)		0.62 ± 0.02 %
Manganese	88.6–95.0 (2)	52–144 (2)		91 ± 4
Mercury	0.149–0.170 (3)	0.10–0.18 (15)		0.155 ± 0.015
Nickel	1.2–1.5 (2)			1.3 ± 0.2
Nitrogen	2.723–2.784 % (2)	2.50–2.86 % (8)		2.76 ± 0.05 %
Phosphorous	0.206–0.211 % (1)	0.14–0.31 % (13)		0.21 ± 0.01 %
Potassium	1.444–1.506 % (2)	1.11–1.62 % (18)		1.47 ± 0.03 %
Rubidium	11.0–12.5 (2)	10.3–12 (3)		12 ± 1
Scandium		0.04–0.2 (3)	0.04	
Selenium	0.076–0.091 (2)	0.08–0.21 (5)		0.08 ± 0.01
Sodium	71–90 (3)	40–524 (12)		82 ± 6
Strontium		23–45 (6)	37	
Uranium	0.026–0.031			0.029 ± 0.005
Zinc	22.8–28.2 (3)	18–81 (25)		25 ± 3

* Concentrations given in µg/g of dry sample unless otherwise specified.
** Range given is of individual values reported for NBS results and of mean values reported for other laboratories. Number in parentheses is for the total number of groups reporting values.
• Values reported for arsenic cluster around concentrations of 14 µg/g and 10 µg/g. At the present time a value of 10 µg/g should not be considered erroneous.

concept of standardized methods has never developed a strong constituency.

Standard reference materials of use to marine pollution studies have been prepared by the United States National Bureau of Standards. At present only values for inorganic substances are available. The materials available are orchard leaves and bovine liver, while a third material, albacore tuna muscle, is in preparation. Preparation of such standards usually takes several years. The heavy metal concentrations of these materials span the ranges of those usually encountered in marine organisms. For the orchard leaves there are nineteen certified values for metals, and for bovine liver, twelve (Tables 46 and 47). The materials are available at modest cost from the National Bureau of Standards, Washington, D.C. Primary standards for heavy metals in sedimentary materials are described in the tabulation of Flanagan (1974).

The varying abilities of laboratories to cope with the difficulties of performing assays upon trace pollutants in marine samples are illustrated in two recent evaluations. The first involved the

TABLE 47. Values reported for bovine liver (NBS-SRM 1577)*

Element	Range of results**		Recommended interim value	NBS certified value
	NBS	Other		
Antimony		<0.1–.014 (2)	0.014	
Arsenic	0.050–0.059 (1)		0.055	
Cadmium	0.24–0.32 (3)	<0.1–0.35 (6)		0.27±0.04
Calcium	117–125 (1)		123	
Cesium		0.013 (1)	0.013	
Chlorine	2,533–2,656 (1)		2,610	
Chromium		0.18 (1)	0.2	
Cobalt	0.17–0.19	0.15–0.29 (3)	0.18	
Copper	181–204 (3)	120–202		193±10
Iron	259–299 (2)	232–280 (6)		270±20
Lead	0.26–0.41 (2)	0.33–2.5 (3)		0.34±0.08
Magnesium	600–611 (1)		605	
Manganese	9.3–11.3 (2)	10.4–27.9 (7)		10.3±1.0
Mercury	0.014–0.017 (2)	0.006–0.070 (7)		0.016±0.002
Molybdenum	3.08–3.34 (1)		3.2	
Nitrogen	10.25–11.03 % (2)			10.6±0.4 %
Potassium	0.93–1.01 % (2)	0.756–0.99 % (3)		0.97±0.06 %
Rubidium	17.5–19.2 (2)	16 (1)		18.3±1.0
Selenium	1.05–1.17 (2)	0.95 (1)		1.1±0.1
Silver	0.054–0.081 (1)	<0.05–0.11 (2)	0.06	
Sodium	0.223–0.267 % (2)	0.20–0.267 % (3)		0.243±0.011 %
Strontium	0.136–0.142 (1)		0.14	
Uranium	0.00055–0.00090 (1)		0.0008	
Zinc	120–134 (3)	78.5–242 (10)		130±10

* Concentrations given in $\mu g/g$ of dry tissue unless otherwise specified.
** Range given is of individual values reported for NBS results and of mean values reported for other laboratories. Number in parentheses is for total number of groups reporting values.

standardization techniques for an international co-operative study of organochlorine and mercury residues in wildlife sponsored by the Organization for Economic Co-operation and Development (Holden, 1973). Twenty-six laboratories in twelve countries participated. Three test samples for organochlorine materials were prepared for circulation among the laboratories: one organochlorine sample contained seven pesticides or derivatives added to a corn oil; a second was a standard solution of a PCB formulation in hexane; and the third a homogenate of cormorant muscle containing mainly PCBs. The test samples containing mercury consisted of a group of freeze-dried homogenates of the muscle tissue of pike and a solution of methylmercury dicyandiamide in water. The samples were analysed both for total mercury and for methyl mercury.

The analytical techniques for organochlorine analysis, although all involved the gas chromatograph, differed in methods of extraction, column packing, clean-up techniques, etc. For the corn oil, most laboratories reported six of the compounds (heptachlor epoxide, p,p′ DDE, p,p′ DDD, p,p′ DDT, o,p′ DDT, and dieldrin), but were unable to identify the seventh, dichlorobenzophenone. The coefficients of variation for the six compounds ranged between ±10 per cent and ±15 per cent.

The PCB solution contained 9.8 mg/l. Twelve results varied between 8.4 and 12.6 mg/l with a mean of 10.2 and a coefficient of variation of 10.2 per cent. Different PCB standards were used by the participating laboratories. Holden (1973) concluded that the agreement among the analyses was very satisfactory.

Fourteen laboratories reported on the PCB content of the cormorant tissue with an over-all range from 240 to 525 mg/kg. Eight laboratories that used the same PCB standard submitted values between 279 and 462 mg/kg with a mean of 374 mg/kg.

Nine laboratories participated in the analyses of the fish tissue for mercury and/or methyl mercury. One laboratory determined total mercury by neutron activation analysis while the others utilized flameless atomic absorption techniques. The Euratom laboratory reported 0.55 mg/kg while the others obtained values between 0.29 and 0.36 mg/kg. The true value was 0.35 mg/kg. The results for methyl mercury were in good agreement, although less than those for total mercury. All laboratories used electron-capture gas chromatography in the detection step. However, the extraction and clean-up steps varied from one laboratory to another. The true value for methyl mercury was given as 0.35 mg/kg, while the reported values ranged between 0.27 and 0.46 mg/l.

This exercise established that it is possible to achieve good agreement among laboratories using different methods in the assay of organochlorines and mercury. It must be emphasized that the work in the laboratories was supervised by experienced analysts.

A somewhat different picture of the analytical abilities of environmental chemists emerged from an interlaboratory investigation of lead in sea water (Anon, 1974). Nine laboratories participated in the study. The analysts utilized atomic absorption and anodic stripping voltametry techniques, which are relatively rapid and inexpensive, and isotope dilution which is too slow and expensive for routine applications. The latter technique was assumed to give accurate results.

Great care was exercised in the preparation of the samples to ensure that all participants analysed a solution containing the same amount of lead. Analyses were carried out during the same time period (within a month of each other and none at a period longer than two months after collection). The sea water taken off the coast of California was unusually clear and there

TABLE 48. Lead concentrations in La Jolla sea-water samples collected 1 November 1974 and analysed by anodic stripping (AS), atomic absorption (AA) and isotope dilution (ID)

Partici-pating laboratory	Method	Amount per analysis (g)	Analysis date	Analytical results (ng/kg)	Best value
1	AS	10	20.XI.1972	Mean of 8 values	1,300
2	AA	500	20.XI.1972	50, 50	50
3	AA	1,220	21.XII.1972	120, 60	60
4	AS	15	2.XII.1972	Mean of 3 values	600
5	AA	700	25.XI.1972	55	55
6	AS	90	5.XII.1972	200, 180, 460, 450, 120, 180, 170	180
7	AA	500	1.XII.1972	120	120
8				Did not report	
9	ID	1,000	20.XI.1972	Standardized value	14±3

was no significant amount of lead associated with particulate phases.

The results were appalling. For the sample containing the lowest amount of lead (14 ng/kg as determined by isotope dilution), both atomic absorption and anodic stripping voltametry techniques gave erroneously high results (Table 48). The lead concentrations as measured by atomic absorption methods were 4 to 8 times higher than the standardized value. Anodic stripping voltametry yielded values 10 to 100 times too high. For sea waters, subsequently analysed, with higher lead concentrations the agreement was somewhat better. Material balance calculations indicated that the amount of lead in the blanks run by the two techniques were of the same order as the amount in the sea water. Two remedies to improve the methods of analyses were evident: (a) the use of larger samples coupled with pre-concentration techniques; and (b) the development of cleaner laboratory practices to reduce the procedural blank. It might be pointed out that the coastal waters, utilized in the above experiment, contained half the concentration of lead in any deep water previously analysed by isotope dilution methods. The samples were taken with great care to reduce any introduction of lead during the collection steps. This intercomparison activity clearly demonstrated the difficulties of even experienced analysts in accurately measuring very small amounts of lead in sea water.

For other sea-water pollutants for which primary standards do not exist or for which interlaboratory comparison experiments have not been carried out, the validity of the reported analyses may be as satisfying as that of the halogenated organics and mercury in biological materials, or as unsatisfying as that of lead in sea water. Perhaps the solution of the problem of the occurrence of strontium-90 in deep oceanic waters (see Chapter 4) would have been found if standard reference materials had been available or if interlaboratory exercises had been carried out. As it now stands, the accuracy of analyses of either of the two laboratories remains in doubt.

Effective monitoring programmes demand accurate analytical procedures.

Bibliography

ANON. 1974. Interlaboratory lead analyses of standardized samples of seawater. *Marine Chemistry*, vol. 2, p. 69–84.

BERNER, L.; MARTIN, J. H.; McGOWAN, J.; TEAL, J. 1972. Sampling marine organisms. In: E. D. GOLDBERG (ed.), *Marine pollution monitoring: strategies for a national program*, p. 11–16.

FLANAGAN, F. J. 1974. Reference samples for the earth sciences. *Geochim. Cosmochim. Acta.* vol. 38, p. 1731–44.

GOLDBERG, E. D. *et al.* 1971. Proposed baseline sampling program. In: *Chlorinated hydrocarbons in the marine environment*, p. 23–36. Washington, D.C., National Academy of Science.

HOLDEN, A. V. 1973. International cooperative study of organochlorine and mercury residues in wildlife, 1969–71. *Pestic. Monit. Bull.*, vol. 7, p. 37–52.

PRESTON, A.; WOOD, P. C. 1971. Monitoring the marine environment. *Proc. Roy. Soc. London*, vol. 177B, p. 451–62.

PRESTON, A.; MITCHELL, N. T. 1973. Evaluation of public radiation exposure from the controlled marine disposal of radioactive waste (with special reference to the United Kingdom). In: *Radioactive contamination of the marine environment*, p. 575–93. Vienna, IAEA.

10.   Health of the oceans

Have the oceans been badly damaged by man's deliberate or unintentional discards of materials into it? Are future threats to their well-being identifiable on the basis of what we know today? The answers to these questions depend upon the period of time over which the assessment is to be made. For both scientific and managerial purposes it is convenient to place ocean pollution problems in two general classes: the short time-scale/coastal-ocean problems and the long time-scale/open-ocean problems. An arbitrary time division between these classes might be a century.

Where should our concerns centre? What types of information are needed to diagnose the ocean's state of health? What actions should we be taking today? These are the questions this final chapter will consider.

Short time-scale problems of the coastal ocean

The evident damage to ocean resources has taken place in coastal waters, usually unpredictably. Such surprises as the Minamata Bay incident and the decline of the Anacapa pelican population were beyond the predictive capabilities of environmental scientists at the times of their occurrences. The respective roles of methyl mercury and DDT in neurological diseases and egg-shell thinning were discovered only as a consequence of tragic events. The lack of concern in the past about the discharge of highly toxic wastes, such as those containing mercury and DDT, to the oceans can be attributed to the then prevailing mood that the oceans were essentially a bottomless receptacle. Today such a concept is totally unacceptable. The detailed statistics on what is entering the coastal ocean are essential to a nation desirous of maintaining its marine resources. It is probably safe to predict that more unpleasant surprises will develop regarding pollution in coastal waters.

The protection of their coastal zones is one of the housekeeping problems of the coastal nations. A nation's environmental scientists must continually assess the quantity of materials its society is discharging to the oceans. As a coastal nation enters the nuclear age, the management of its radioactive wastes becomes a first-order problem. A portion will enter the coastal zone. Will consumers of sea foods or those using the beaches for recreational purposes be exposed to unacceptable levels of ionizing radiation? The siting of nuclear-power plants, nuclear weapons arsenals and nuclear reprocessing facilities requires inputs from the marine community who can assist in formulating risk-benefit alternatives for the estimated releases of radio-activity to the oceans. There are numerous examples from the recent establishment of nuclear-power-generating facilities in coastal zones to illustrate that conflicts in use of the area can be eliminated rather than adjudicated (Chapter 5).

Some of the present-day concerns will be tomorrow's irrelevancies. It is probable that the petroleum supplies of the world will be exhausted before effective measures to substantially reduce leakages to the oceans are instituted. In the future the rising costs of heavy metals will cause greater re-use and less loss to the environment than is now being experienced.

Long time-scale problems of the open ocean

The slow but continuous alteration of the open-ocean waters can offer future generations the legacy of a poisonous ocean. It is most unreasonable to titrate the seas with man's wastes to the endpoint of a world-wide mass mortality of organisms. Yet, such an event is today not inconceivable. The time might be a century or longer. Today, we are adding annually megaton quantities of synthetic halogenated hydrocarbons to the ocean system. Surface-water values today are of the order of nanograms per litre (see Chapter 3). If these substances follow the water in mixing with the deep ocean, they will be transferred within a decade to zones below the mixed layer, where they may remain for thousands of years, the residence time of the persistent naturally occurring organic molecules. At what level might they irreversibly damage the ecosystem?

It is concerns of this type (and others can be orchestrated in similar detail) that are within the provenance of international organizations. The potential pollution of the open ocean will result from the contributions of many nations, all of whom have some economic stake in the loss or restricted use of these resources. Yet, it appears that the economic and scientific resources of any single nation are insufficient to engage in appropriate and adequate surveillance activities concerning the state of the ocean's health.

Strategies of diagnosis

A continuing diagnosis of the present state of health of the open oceans and predictions for the future require access to the following information: (a) production and use data, including the locations of production and use, for materials as well as fuels for the generation of energy; (b) the physical and chemical properties of the substances of concern; and (c) levels of toxicity and exposure that offer an acceptable risk. The mechanics to obtain such information should be developed by international agencies, after the specific data sought are provided by the environmental science community. The tactics for compiling this information will vary. For example, the production and use data on halogenated hydrocarbons requires a mail and telephone programme of queries, once the release of such data is approved by the world's nations. On the other hand, the formulation of acceptable exposure levels for such compounds as hexachlorobenzene or curium will require extensive investigations in modern, high-quality laboratories.

PRODUCTION AND USE DATA

Although there are many thousands of chemical compounds manufactured or mined by man, only a very small percentage are of concern to the environmental scientist. For potential pollution of the open ocean, in general only those substances that are produced in annual quantities of more than 1,000 tons (excluding of course the radio-active nuclides) are necessary for assessment (see, for example, the methods used in the United States National Academy of Science Report *Assessing Potential Ocean Pollutants*, 1975).

PHYSICAL AND CHEMICAL PROPERTIES

In order to predict environmental behaviours, certain physical and chemical properties of a substance must be known. They may include such measurements as vapour pressure, stability under exposure to ultraviolet and visible light, and ease of microbial degradation. There are few laboratories in the world today with the mission and/or abilities to carry out the variety of measurements necessary. An international organization could play a most substantial role in developing methods to obtain such information.

TOXICITY AND RISK DATA

Information on toxicity and acceptable risk will be the most difficult of all to tabulate. Although recommended environmental levels for the artificial radio-active nuclides have been established based upon public health considerations, there is little substantial information for other chemicals. An attempt to define acceptable levels of mercury in fish has been made (see Chapter 1), on the basis of scanty information. There is some controversy over acceptable atmospheric burdens of plutonium. The question as to whether there is a threshold limit for a pollutant exposure remains unresolved for the case of ionizing radiation, and for other substances the question is often not even posed. The development of venturesome and innovative approaches can become the concern of an international agency.

Strategies of treatment

The collation of the above information for potential pollutants will yield a set of priorities for the management of materials. Models can be constructed for those substances whose build-up in the coastal or the open ocean can result in unacceptable risks. What are the tactics for the protection of the ocean—its coastal water by individual nations and its open-ocean waters by international agreements?

The goal of management strategies is the maintenance of the widest possible margin between the acceptable level of a pollutant in ocean waters and the measured amount. Knowledge of environmental levels are necessary. Since the cost of surveillance programmes is high, every effort should be made to introduce economy by minimizing the number of actual measurements. Where there are wide gaps between acceptable levels, measurements can be made at longer intervals in fewer places.

In the open ocean, measurements of the extremely low concentrations of pollutants in deep waters will be most demanding upon the skills of the chemist. Past history has indicated that extreme care must be taken to avoid contamination. The unresolved problem of strontium-90 in deep ocean waters and the inadequacies of lead analysis serve as cautionary guides for analyses of other low-level pollutants.

The difficulties of scientific problems are comparable to those of a political nature. When a pollutant source is identified which leads or can lead to unacceptable levels in the open ocean, what types of action can be taken to reduce its emissions? The recognition by both scientists and governments that the open ocean is vulnerable to poisoning by slow but consistent pollutant leaks is the foremost and fundamental step.

The mussel/barnacle watch—a first step

The many proposed global marine monitoring programmes are characterized by their vastness and complexity which lead to their doom as fantasies on paper. Inputs from biologists, chemists, physicists, geologists, meteorologists and engineers have indicated a need for measurements which would tax the facilities of the existing world marine science community. While such documents pass for review from one international organization to another, the world ocean continues to receive man's wastes, and there is no systematic attempt to measure the exposure levels of identified pollutants in the various parts of the ocean.

The pollutants of the marine system so far identified have concentrations in the parts per billion level and less in marine waters, and are extremely difficult to measure. Talented analysts using highly sophisticated techniques are necessary to achieve reliable concentrations of these substances in materials from the marine environment. For some (the halogenated hydrocarbons and petroleum hydrocarbons) measurements in sea water are so taxing, due to the extremely low concentrations of the pollutants, that only a few dozen reliable measurements have been performed. As a consequence, sea-water assays, instead of the use of sentinel organisms, may not be the most reasonable way to ascertain marine concentrations of these sets of pollutants.

Mussels and barnacles are especially attractive as sentinel organisms for the purpose of measuring the exposure levels of these pollutants. The mussel *Mytilus edulis* has been extensively studied both experimentally and ecologically with respect to its ability to record pollutant levels. Plutonium exposure levels appear to be recorded in both the shell and in the soft parts (Noshkin *et al.*, 1973). Molluscs are well-known concentrators of heavy metals (Lowman *et al.*, 1971). Mussels have been used already in a regional monitoring programme of halogenated hydrocarbons and mercury (Holden, 1973). The mussels appear valuable as sentinels for hydrocarbon pollution. They rapidly take up both saturates and aromatics from their environment and store them with little metabolic breakdown (Lee *et al.*, 1972). *Mytilus edulis* rapidly responds to the total hydrocarbon burden of its environment through uptake in its tissues and also rapidly releases such pollutants upon exposure to clean waters (Disalvo *et al.*, 1975). The species is wide-ranging in the northern hemisphere and there should be no problem about introducing it to places where it may not already occur. Transplantation to areas where the organism is not indigenous has been carried out in San Francisco Bay and off the coast of southern California. The organism is a bay and sheltered coastal species and will record environmental levels well in such localities. It is a filter feeder and its viscera are compact and easily separable from the shell.

A second group of organisms which appears worthy as sentinel organisms are the goose barnacles (*Lepas*). Unlike the mussel it is an oceanic species, easily caught in neuston nets. Although the 20- or 30-odd species that make up the genus are sometimes difficult to identify, the species inhabiting one region should be relatively unchanging with time. The goose barnacle is quite cosmopolitan, found in nearly all coastal waters. Less work has been carried out on its body burden of pollutants than on that of the mussel, *Mytilus*, yet it is now attracting increased attention as a sentinel organism.

With this background we can proceed to estimate the annual costs of a global monitoring system. Clearly, the scope of the programme will determine the resources needed. Perhaps more important is the determination of the amount of information that could be considered adequate for assessing the present health of the waters and the trends in pollutant levels. For the moment we will assume that an initial run of 100 samples in ocean waters not under the jurisdiction of sovereign nations would provide an initial entry to the surveillance programme.

The analytical costs in Table 49 were obtained from practising analysts.

The pollutants to be measured include those substances that have been shown to jeopardize life processes in the ocean or that are similar to chemicals that have been associated with undesirable impacts upon living systems: heavy metals (lead, cadmium, mercury, selenium, zinc, silver and copper as total concentrations); chlorinated hydrocarbons (DDT residues, polychlorinated biphenyls, hexachlorobenzene, dieldrin, endrin, heptachlor, benzene hexachloride, cis- and trans-chlordane and the insecticidal derivatives oxychlordane and

TABLE 49. Analytical costs,* based on analyses of 100 samples, including intercalibration exercises, and the analyses of standard samples

Analyses	Cost ($)
Petroleum hydrocarbons at $860 per sample	86,000
Artificial radionuclides at $300 per sample	30,000
Chlorinated hydrocarbons at $100 per sample	10,000
Heavy metals at $20 per sample	2,000
TOTAL	128,000

* These costs are for the time of the analyst and necessary chemicals. They do not include any cost for capital equipment or overhead.

heptachlor epoxide); artificial radioactive nuclides (plutonium-238, plutonium-239, 240, americium-241 and cesium-137); and petroleum hydrocarbons (to include measures of the concentrations of alkanes, cycloalkanes and aromatics including 2-, 3-, 4- and 5-ring polynuclear condensed species).

The costs of collection, shipment, storage and distribution of the samples would probably be a figure of the same order of magnitude. Thus, the entire programme could be underwritten for less than $300,000.

Sampling sites could be selected upon the basis of a rational geographic distribution and upon accessibility. Since the concerns will be with open-ocean pollution, it will be desirable to stay away from major continental pollution sources.

Sampling intervals might be annual. Where a given suspected pollutant becomes too low in concentration to measure, or where it approaches the levels of uncontaminated areas, a cessation of its assay appears reasonable. Splits of all samples should be maintained in a freezer library as reference materials for future baseline studies or for analyses of subsequently identified pollutants.

Bibliography

DISALVO, L. H.; GUARD, H. E.; HUNTER, L. 1975. Tissue hydrocarbon burden of mussels as potential monitor of environmental hydrocarbon insult. *Environ. Sci. Technol.*, vol. 9, p. 247–51.

HOLDEN, A. V. 1973. International cooperative study of organochlorine and mercury residues in wildlife, 1969–71. *Pestic. Monitor. Bull.*, vol. 7, p. 37–52.

LEE, R. F.; SAUERHEBER, R.; BENSON, A. A. 1972. Petroleum hydrocarbons: uptake and discharge by the marine mussel *Mytilus edulis*. *Science*, vol. 177, p. 344–6.

LOWMAN, F. G.; RICE, T. R.; RICHARDS, A. F. 1971. Accumulation and redistribution of radionuclides by marine organisms. In: *Radioactivity in the marine environment*, p. 161–99, Washington, D.C., National Academy of Science.

NAS. 1975. *Assessing potential ocean pollutants*. Washington, D.C., National Academy of Science. 438 p.

NOSHKIN, V. E.; BOWEN, V.T.; WONG, K. M.; BURKE, J. C. 1973. Plutonium in North Atlantic Ocean organisms: ecological relationships. In: D. J. NELSON (ed.), *Radionuclides in ecosystems, Proc. Third Nat. Symp. Radio. Ecology, May 10–12, 1971, Oak Ridge, Tenn.* Vol. 2, p. 681–8.

Definitions and abbreviations

This list contains the definitions of those abbreviations of measurements used in the text that might not be familiar to the non-scientific reader.

DIMENSIONS

m², m³	Square metres, cubic metres.
km², km³	Square kilometres, cubic kilometres.
μm	Micrometre or micron (1/1,000,000 of a metre).

MASS

ng	Nanogram (1/1,000,000,000 of a gramme).
μg	Microgram (1/1,000,000 of a gramme).
mg	Milligram (1/1,000 of a gramme).

RADIO-ACTIVITY

dpm	Disintegrations per minute.
Ci	Curie, or approximately 3.700×10^{10} disintegrations per second. About that of 1 gramme of radium.
mCi	MicroCurie.
half-life	The time required for half the atoms in a given quantity of a radio-active element to disintegrate.

PHYSICAL CHEMISTRY

pH	The logarithm (base 10) of the reciprocal of the hydrogen ion concentration in gramme-molecules per litre; measure of acidity, neutrality and alkalinity.

POWERS, PREFIXES AND RATIOS

$10^1, 10^2, 10^3$, etc.	10, 100, 1,000, etc.
$10^{-1}, 10^{-2}, 10^{-3}$, etc.	1/10, 1/100, 1/1,000, etc.
k	Kilo-, or 1,000 times (another quantity or measure).
p	Pico-, or 1 million millionth (of another quantity or measure); $10^{-12}x$. (See also 'Dimensions and Mass'.)

p.p.m.	Parts per million.
p.p.b.	Parts per billion.
p.p.t.	Parts per trillion.

MISCELLANEOUS

<	Is less than.
>	Is more than.